Dolly Mixtures

THE REMAKING OF GENEALOGY

Sarah Franklin

Duke University Press Durham and London 2007

Printed in the United States of America
on acid-free paper ∞
Designed by C. H. Westmoreland
Typeset in Cycles with Euphemia display
by Tseng Information Systems, Inc.
Library of Congress Cataloging-in-Publication
Data appear on the last printed page
of this book.

Dolly Mixtures

A John Hope Franklin Center Book

TO MARTHA POTTER BROOKS,

"TYPPI" (1912–99), for being

a rare breed

Contents

Acknowledgments

This project began as a Leverhulme Trust Research Fellowship in 2000, for which I am very grateful. I am also indebted to Ian Wilmut and his colleagues at the Roslin Institute, which I was able to visit in 2000 and again in 2001 for highly educational research tours of the facility. There, I also met Dolly and her many flock-mates, who were as charmingly intractable as I had heard. The opportunity to spend time with Thelma Rowell and her flock of feral Soay sheep in Yorkshire, near Ingleton, was enhanced by many conversations and homemade delicacies that provided me with insight into many aspects of animal sociality of which I was unaware, as well as aspects of my own I had never noticed. I am very indebted to Thelma for her generosity, hospitality, and wisdom about sheep in particular, and life in general. Thank you also to Glen and Dan Shapiro for an introduction to your Blue-faced Leicester flock on Hazelwood Farm and many conversations about "sheepwatching." Lastly, I owe considerable thanks to the Guy's, King's and St. Thomas's Stem Cell Consortium, in particular Peter Braude, Sue Pickering, and Stephen Minger, who have enabled me to learn from them about stem cell science, embryology, and developmental biology.

The number of individuals who have assisted in the intellectual journey that accompanied the writing of *Dolly Mixtures* is almost as vast as the number of people who should be thanked for putting up with endless sheep jokes. In Britain, I owe a big thanks to my former Lancaster colleagues, especially Alan Holland, Maureen McNeil, and Jackie Stacey, as well as my BIOS Centre colleagues at the London School of Economics and Political Science, in particular Nikolas Rose. In the United States, I have to thank Lauren Berlant for leading me to the sheep entry in the *Encyclopedia Britannica*, and Charis Thompson for inviting me to visit the University of Illinois at Urbana-Champaign, where Elizabeth Franklin helped to remind me of the depth of my own agricultural origins. In Australia I would particularly like to thank Sue Hawes for hosting my visit to Monash, and Elspeth Probyn, whose hospitality at the University of Sydney enabled the writing up of the final draft of *Dolly Mixtures*. In Canada I would like especially to thank Bruce Greenfield for sending me John T. Juricek's critique of Frederick Jackson Turner's frontier hypothesis, which, crossbred with Patrick Wolfe's concept of the frontier effect (thank you Fiona Probyn!), assisted greatly with the rethinking of genealogy. Jill Ker Conway, who hails from all of these places, generously assisted me by providing copies of her early work on sheep economies in colonial New South Wales, as well as the inspiration of her own autobiographical writings about her upbringing in the Australian outback.

The work of many scholars has been prominent in my thinking throughout the writing of *Dolly Mixtures*, in particular that of Rebecca Cassidy, Donna Haraway, Gillian Feeley-Harnik, Molly Mullin, Harriet Ritvo, and Marilyn Strathern. I am also very grateful to everyone who participated in the two panels Molly Mullin and I organized at the American Anthropological Association meetings titled "Anthropology's Animals" and "The Animal Turn." These proved formative events in my thinking, and they continue to be a source of inspiration.

To everyone who sent me sheep-abilia, ewe know who ewe are! Thank you and may this form of ovine reproduction never cease. To all of the audiences who heard versions and portions of *Dolly Mixtures*, I am very indebted to your comments, criticisms, and suggestions. To the readers at Duke University Press, and the editorial and production staff, I am very indebted to your ongoing assistance, encouragement, and advice.

Above all I would like to thank Sara Ahmed for her contribution to thinking through genealogy as an "orientation"—it is a pleasure and a privilege to benefit from her insights and suggestions. This book is dedicated to Martha Potter Brooks, who early on steered me in the right directions and taught me how to use all of my senses. It is from her I learned that curiosity is an emotion, for which I have always been particularly grateful.

Origins

> Mice and humans in technoscience share too many genes, too many worksites, too much history, too much of the future not to be locked in a familial embrace.
> —Donna Haraway, *Modest_Witness@Second_Millennium*

> Whether we assent or not, the laboratory mouse's very existence engages us in a complicated process of technical, cultural, and political formation, but at the same time, its historical situatedness provides a tool kit for intervening in this process.
> —Karen Rader, *Making Mice*

In various speaking engagements throughout the writing of this book I often encountered audiences that were somewhat doubtful about how much there was left to say about the topic of cloning or the birth of Dolly the sheep. The association of cloning with repetition and endless similarity seemed to be replicated in their anticipations that the topic was overrehearsed and now somewhat dated. This book challenges these doubtful anticipations by arguing that we have only really just begun to develop a suitable critical language for parsing the significance of Dolly's coming into being. It is as though she is hidden behind a hedge of bad puns in a paddock. The argument presented here differs primarily from other accounts of cloning in that it does not take a legal, bioethical, theological, or public policy approach,[1] but instead, building on the work of historians and anthropologists, tries to situate her emergence as part of the history of agricultural innovation and its close connections to the life sciences—in particular reproductive biomedicine. From this perspective, Dolly's genealogy is thick with significance—as a form of animal capital, as a very British animal, and as an animal model of a technique that has significant potential to improve human health.

Dolly Mixtures are a British type of candy, or sweet, made up of brightly colored different pieces marked with distinctive stripes and patterns. Like their kindred confection "all sorts," Dolly Mixtures are

Dolly Mixtures are a smaller version of an earlier British candy known as Empire Mixtures made up of multi-colored pieces of marzipan. *Photograph by Sarah Franklin.*

a popular treat and a staple commodity at newsagents, grocery shops, corner stores, cinemas, and supermarkets alike. They are a distinctive species of confection, made up of an assortment of ingredients, and presented as a variety of shapes. At once a distinctive type and a blend of components, Dolly Mixtures seemed a fitting point of departure to think about Dolly as a uniquely blended British breed.

From this perspective Dolly is a mixture not only because she embodies a novel technique for combining genes and cells but because she constitutes the outcome of a lengthy and complex historical and biological genealogy as an experimentally bred sheep. It is consequently unclear where Dolly's genealogy begins and ends. However, what is apparent is that a broad cultural and historical analysis of her coming into being widens the significance of her birth beyond the overprominent question of whether or not humans should be cloned. While Ian Wilmut's claim that Dolly is "the most extraordinary creature ever born" may be dismissed as hyperbolic, exaggerated, or even self-promotional, this book argues that it cannot simply be ignored as pure hype. To the contrary, the importance of Dolly's birth—as the founder animal not only of a new form of reproduction (transgenesis) but for a novel realignment of the biological, cultural, political, and economic relations that connect humans, animals, technologies, markets, and knowledges—needs to be situated in a world historical frame.

Dolly thus presents an opportunity to develop conceptual, descriptive, and analytical languages to identify and evaluate the future reproductive possibilities post-Dolly biology now faces, and at the same time to look back at some of the important historical formations that enabled and prefigured Dolly's creation. The use of a genealogical perspective on Dolly assists in framing the questions she poses in terms of *orientation*—where did she come from, and what does her making point toward? If she is a biological frontier, what does this mean, what kind of frontier is she, and how does the idiom of the frontier shape our relationship to the idea of biology?

Dolly not only embodies the legacies of embryology and reproductive biology but of selective breeding and the industrialization of livestock through pastoralism—a process central to the emergence of peoples, nations, colonies, and capital, and now—more evidently than before, but not for the first time, to the life sciences. Dolly also embodies the long history of animal domestication, as well as the more recent histories of capital accumulation through selective breeding and enhanced national competitiveness through control of the germplasm—which can themselves be traced both to biblical themes (Jacob and his sheep) and to Neolithic innovations (companion animals). This book argues that locating Dolly within these longer histories, including the history of industrialization in England and British imperialism, helps to clarify both her significance as a contemporary individual—a viable offspring—and the dense accumulation of genealogical relations that connect her to the past.

By investigating Dolly's genealogy as a historical trajectory in this way, I argue it is possible to sketch a more complex politics around Dolly's significance as a biocultural entity, using the conflation *biocultural* to emphasize the inseparability of the new biologies from the meaning systems they both reproduce and depend upon, such as beliefs about nature, reproduction, scientific progress, or categories such as gender, sex, and species. The primary idioms I use in this book are those of genealogy and mixtures—in order to explore Dolly's connections and disconnections, their scale and direction, their form and transformation. This project uses Dolly as a focal point to ask how thinking through her body can help identify and characterize changes in definitions of the human, technology, and the future, which are the

Dolly gave birth to six lambs, three of which are pictured here in June 2001. Much of her fame derives from the fact she was a perfectly ordinary sheep, and her ability to breed naturally helped to confirm her fitness. *Courtesy of Roslin Institute.*

broad questions at stake in the effort to think carefully about the increasing prominence of the highly technologized forms of assisted reproduction that are reshaping lives, worlds, knowledges, economies, and imaginaries so publicly, explicitly, and visibly at the beginning of the twenty-first century. In sum, *Dolly Mixtures* asks how we can position a shape-shifting sheep within a broader discussion about kind and type, species and breed, sex and nation, empire and colony, capital and livestock—to all of which categories and identities Dolly's existence adds a transformative element.

Dolly was accurately described to me in an interview with Bill Ritchie, the Roslin Institute embryologist who literally pieced her together under the micromanipulator using handmade pipettes the width of human hairs, as "a twentieth-century icon." As one of the most celebrated animals ever created, Dolly's public profile and image are as tightly wound up with the cultural politics of her existence as her distinctively foreshortened chromosomes. While it is commonplace to attempt to separate the sober scientific facts of Dolly's creation from the hype that surrounds her, and the topic of cloning more broadly, this book argues against the viability of such separations by emphasizing many of the ways in which "hype" structures the speculative futures out of which such animals come into being. Insofar as *hype* refers to imagined sets of connections, and exaggerated implications, it signifies a reaching beyond, or an expansion of range. Dolly is in this sense both a frontier and a horizon—a relational someplace and no place signaling future possibility and direction. Both practically, as an embodiment of a technique, and symbolically, as an iconic animal, Dolly takes us out of bounds and beyond the pale of the normal or ordinary scale of things. The "hype" surrounding Dolly cannot therefore merely be dismissed as irrational excess: what Roslin scientists refer to as "Dollymania" is by definition one of many languages in which her unique significance must be reckoned.

While hype raises the question of scale or range, this book is equally concerned with the quotidian classificatory orders altered by Dolly's existence, including both orderings of the biological—such as gender, species, or breed—and familiar, established conceptual orderings of life itself—most notably through reproduction, inheritance, descent, and death. Because Dolly's assisted creation out of technologically altered cells confirms the viability of new forms of coming into being, or procreation,[2] her existence can be seen to redefine the limits of the biological, with implications for how both sex and reproduction are understood and practiced. Similarly, because cloning has become increasingly interrelated with the topics of immortality and cell death, her existence raises questions about the place of life in the production of pathology and morbidity, and vice versa.[3] Stem cells, and the emergent industries of cell line banking and tissue engineering, integrate life and death in new ways that are both culturally and economically significant—and ask for critical attention.[4] In chapters on both sex and death, this book attempts to widen the debates about Dolly and cloning by asking what it means to contrast "biological control" with its opposite—biology out of control—which is an equally important theme within agriculture, medicine, science, industry, and economics.

Since Dolly is an animal whose making belongs to a long tradition of innovation in the management of life itself as both an economic and national resource, she is a classic mixture of agricultural, scientific, medical, commercial, and industrial ambitions. Hence, while she is very much a late-twentieth-century animal in terms of the precise molecular technologies necessary to her creation, the feat of producing her viability belongs to a long tradition of reshaping animal bodies, crisscrossing cell lines, and redesigning animal germplasm in the interests of both capital accumulation and national or imperial expansion.[5] One need only think briefly about the importance of the sheep and wool markets to the industrial revolution in Britain in the eighteenth century, or of the export of British animals to the new world to produce new markets as part of colonization, to make the connection between Dolly, the British biotechnology industry, and the Australian outback. It is only a small step in the opposite direction to find one's way to IVF (in vitro fertilization) via Australian sheep and thus to embryonic cell lines and contemporary global bioscientific innovation.[6] These pathways are explored in chapters entitled "Capital," "Nation," and "Colony" in *Dolly Mixtures*.

Throughout the account of the remaking of genealogy, to which Dolly is heir, this book draws on a wide and eclectic interdisciplinary range of approaches and sources, from historical accounts of sheep breeding to scientific representations of cloning by nuclear transfer to popular media accounts of Dolly's creation and birth. Critically, this book draws on cultural studies and anthropology, gender and kinship theory, science studies, postcolonial criticism, and the history of biology. At issue is the complicated intersection between genetics and biology as social forms and their embodiment as sets of practices, techniques, and animate entities such as cell lines or cloned sheep. By emphasizing throughout this book that biology is socially produced, thick with specific and accumulated histories, and always already culturally mediated in each situated encounter, the attempt is not to produce a "purely discursive" account of either cloning or Dolly the sheep. By foregrounding the mixtures out of which Dolly is produced, this book attempts instead to explore the integration of biology and genetics as material forms with their production as representational systems, including scientific knowledge and instrumentalism. The biology of the domestic sheep, as many distinguished sheep historians have argued, is inseparable from human history—and vice versa (Ryder 1983). We can say the same of stem cells. That the language of biology and biotechnology can be read as indexical of the classificatory pathways through which objects, fields, processes, and accounts of causality are continually reordered is by now a very well-established principle—both within and outside of science itself

Roslin Institute near Edinburgh in Scotland is one of the world's premiere animal agricultural facilities. It is internationally recognised for its research programs on molecular and quantitative genetics, genomics, early development, reproduction, and animal behaviour and welfare, as well as its cloning and transgenics programs. *Courtesy of Roslin Institute.*

(Haraway 1991, 1997). If the conventions of scholarly writing about such reorderings, or critical analysis of them, are less well recognized or established, few would disagree there is an urgent need for more nuanced and sophisticated responses to the complex intersections of the social and the biological that are literally reshaping reproduction, genealogy, and inheritance.

The "animal turn" in many disciplines that have seen an explosion of scholarship on animal-human connections has provided a vast array of innovative approaches this book draws upon, from the powerfully provocative accounts of primates, mice, and dogs authored by Donna Haraway (1989, 1997, 2003) to the meticulous dissection of animal conundrums in the early period of British industrialization and imperialism provided by Harriet Ritvo in *The Animal Estate* (1987) and *The Platypus and the Mermaid* (1997).[7] As the quotations from Haraway and Karen Rader concerning laboratory mice reproduced as the epigraphs to this chapter emphasize, the remaking of animal genealogies is inextricable from the social values and historical conditions of their human authors. Rebecca Cassidy's brilliant analysis of thoroughbred racehorse bloodlines as crucibles of class, gender, kinship, and Englishness (2002) has directly infused this study, as have earlier anthropological forays into what Roy Willis describes as *Signifying Animals* (1990). In particular, Gillian Feeley Harnik's pigeon and beaver stories (1999, 2001, 2004, and forthcoming) about what is in the bloodlines of early anthropological accounts of genealogy, consanguinity, and kinship led me back to Dolly with a renewed appreciation of her Scottish origins. I also learned from sheep breeders and sheep watchers, such as Thelma Rowell (1991a, 1991b, 1993), to think about human sociality "through the animal" more critically.[8]

Diana Fuss makes a common claim in her introduction to *Human, All Too Human* (1996), an anthology of writing about the boundaries of the human named after Friedrich Nietzsche's rumination on the human spirit, that "the political stakes of this pre-eminently philosophical question today pose themselves with special urgency, as debates over the significance of genetic surgery, virtual reality, reproductive technology, artificial intelligence, and other forms of 'posthuman' reconstruction dramatically disorganize traditional Enlightenment conceptions of the human" (1). Indeed, Fuss goes on to claim that "the question of what it means to be human has never been more difficult—and more contested" (1). The birth of Dolly the sheep in the same year Fuss's claims were published could be seen either to confirm or to contradict such claims, as Dolly is as much an extension of the scientific rationality and imperative of improvement that typify the Enlightenment legacy as a necessary threat to the very essence of humanity—as some prominent commentators have claimed (Fukuyama 2002; Habermas 2003; Kass 1998) and some popular science writers have vividly imagined (McKibben 2003). Queering the pitch of these contestations, *Dolly Mixtures* asks if it is necessary to establish a quantitative frame for the difference Dolly makes. Eschewing both the "is it new" and the "is it ethical" versions of the questions raised by Dolly's creation, a qualitative, interpretive, and comparative approach is substituted, through which it is possible to both acknowledge and analyze the ways in which Dolly extends and transforms practices such as animal breeding, tissue engineering, germinal experimentation, capital accumulation, and medical research.

Like its topic, the method used in *Dolly Mixtures* is somewhat deviant, and like Dolly herself, it has been overfed. As an astute reader of the manuscript noted of its meandering path and stupefying accumulation of ovine detail, the problem with following sheep around is that they get everywhere. The path I took from studying cloning to a book about the making of Dolly the sheep was relatively direct: from there, things went rather more chaotically world historical. However, it is true there are few topics that will not lead back to sheep more quickly than I realized prior to researching and writing this book. This does not excuse the primary shortcoming of a method based on "following sheep around," which is the tendency to tread clumsily into areas much more carefully handled by more disciplined scholars. While not unashamedly experimental, *Dolly Mixtures* wears the heavily matted regret of its inevitable errors—of omission and otherwise—as a sheepish mark of respect for theory-on-the-hoof.

At the end of the day, I was driven to scholar-sheep despite its many problems primarily because I felt its benefits were greater than its failings. Hence, it is highly problematic to examine colonialism from the point of view of sheep. But it is also problematic not to. It may be seen as glib to compare the origin of *klon* in twig, the origin of *stock* in stem, and the origin of *genealogy* in trunk, and objections could be raised that these merely etymological resemblances are superficial, or so broad as to be meaningless, and these are not criticisms to which this book turns a deaf ear. However, there is a cost to ignoring these connections, especially when *genealogy* and *reproduction* can be seen as two of the most important, but vague and undertheorized, terms in contemporary critical thought—uniting as they do the work of so many contemporary critical theorists across so many fields, from postcolonial theory to science studies (c.f. Weinbaum 2004). Consequently, if *Dolly Mixtures* is at times uncultivated and promiscuous, and if it has bad scholarly manners, for which it rightly hangs its head, it nonetheless wants to go headfirst into the thicket because the alternative is either to turn around or be penned in—and besides, there might be food on the other side.

One of the most complicated hybridities this book seeks to chart is the relationship between genealogy and social order: after all, both are heavily discredited concepts, burdened with unwelcome historical baggage. Wary of the tendency with which I am familiar from the reconfiguration of kinship studies to keep reinventing the familiar lost object in pursuit of its defamiliarization, I have waded into genealogical territory knowing it might better have been jettisoned for a more contemporary approach to Dolly as a "multiple" (Mol 2002), a "becoming animal" (Deleuze and Guattari 1987), a Lyotardian "inhuman" (1991), or a Latourian "asymmetry" (1987, 1993). Similarly, I have retained the concept of social order despite its having given way to models of the "event" or "assemblage" (Rabinow 1996b, 1999), "mobility" (Urry 1999), "complexity" (Law and Mol 2002) or "flow" (Appadurai 1996). While influenced by these postsocial perspectives, I have retained the idea of sociality linked to older models of social, economic, and biological *order* primarily as a means of emphasizing, or even *over*emphasizing, the ways in which Dolly is socially and economically consequential not only because she is "made to order" but because these "orders" are structural. They may also be disorderly and contingent, contradictory and situated, and as local and individual as Dolly is. However, my argument is that the forms of order that made Dolly possible are organized in ways that have profoundly shaped the past—and will have effects on the future that are, in part, discernible from these histories. This is also why I have borrowed the in some ways very conservative idiom of genealogy—precisely because it makes us look back, and within, to understand the social consequences of the forward and toward that Dolly embodies, and which I describe as her genealogical orientation.[9]

A final "scapegoat" *Dolly Mixtures* wants to engage in a different kind of conversation is the oft-encountered view that in the very process of writing about high-profile, "big" science, the researcher is not only collaborating with its overvaluation through promoting its importance but is also complicit with science in being overawed by its magnificence or potency. This view is often accompanied by a version of the suspicion that the researcher's critical ability is compromised by proximity to the busy scene of scientific innovation that should be kept at a more critical distance, and that working closely with scientists inevitably makes one sympathetic to their cause. Hence, in taking scientific innovation seriously enough to attempt to describe it carefully, there is the risk of being seen to reify, or even celebrate, the value of scientific progress, the heroism of its pursuit, or its overvaunted novelty—thus becoming one of science's apologists.

This book could easily be read in such a manner, as it is neither a denunciation of cloning, nor even a mild critique. It is a book, moreover, that describes as one of its main aims to respond to Ian Wilmut's suggestion that more critical attention needs to be paid to the meaning of what he calls "the age of biological control" (Wilmut, Campbell, and

Ian Wilmut and Dolly. Wilmut's claim that the birth of Dolly has ushered in "the age of biological control" was intended less as a boast than as a warning that the increasing ability to re-engineer life forms poses an ever greater social challenge to set the limits biology "itself" no longer provides. *Courtesy of Roslin Institute.*

Tudge 2000). *Dolly Mixtures* does not avoid a "position" on cloning because it is "pro-science," but because the "pro-" and "anti-" science positions overprivilege scientific innovation as the necessary site of interrogation, impeding more substantial critical engagement with the wider social practices from which "science" is inseparable. *Dolly Mixtures* does not argue the industrialization of livestock of which Dolly is a viable offspring means she is a remnant ruminant of British imperial aspirations, because that is far too simplistic a genealogy. However, it is not irrelevant either that she was born in Britain, home to both imperial and industrial aspirations with which she has strong family ties. It is in the interstices of these ties that *Dolly Mixtures* seeks to relocate the cloning question, which is disproportionately imagined in the future, at the expense of how much there is to be learned about the logic of Dolly's creation in the past.

At a more practical, immediate, and experimental level, a primary aim of *Dolly Mixtures* is to give a good, hard shove to the assumption that important scientific questions can only be properly addressed from within, based on a comprehensive appreciation of the "basic facts" and professional training in the degree of scientific objectivity necessary to rely solely on established ("accepted") factual evidence. While admitting that a highly complex technique such as cloning by nuclear transfer poses an "impossible" challenge for the scientifically untutored commentator—who will inevitably make errors of interpretation and lack the expertise needed either to read or write scientific descriptions as a specialist scientist would—*such failures are nonetheless necessary* to preserve a space of dialogue—and hopefully to widen it.

The space *Dolly Mixtures* seeks to occupy is thus, in another sense, that between popular scientific writing by scientists—such as Ian Wilmut and Keith Campbell, whose account of Dolly coauthored with the science journalist Colin Tudge is a model of scientific translation into accessible language—and that of social scientists who write about science—such as Paul Rabinow (1996a, 1996b, 1999), whose lucid descriptions of scientific knowledge production open a door into its personal, social, and ethical complexity, or Margaret Lock (2002), whose ethnography of organ transplantation is exemplary of what social scientific analysis can provide. Both of these genres create greater accessibility to the questions posed by science by changing the agency of this question to become "what is asked of science"—a question that must be evaluated by scientists and nonscientists alike.

A central theme of *Dolly Mixtures* is the extent to which Dolly can be seen as the product of a mix of well-established scientific questions (Can a higher vertebrate be cloned? Can cellular time be reversed?), commercial questions (Can sheep manufacture valuable human proteins in their milk?), medical questions (Can replacement tissue be grown in a petri dish?), agricultural questions (Can elite animals be genetically duplicated?), industrial questions (Can the energy of cells be harnessed for manufacturing?), and social questions (Will public health benefit from the Dolly technique?). By attempting to depict the long and steady trajectory of such questions that predates Dolly's birth, the question of her origins can be widened into both a more encompassing frame and a more diverse set of theoretical approaches. This also helps to counter the overconcentration on Dolly as a "clone."

This is not to dismiss the huge public response to cloning as an "overreaction"—a term that is commonly misused to describe a

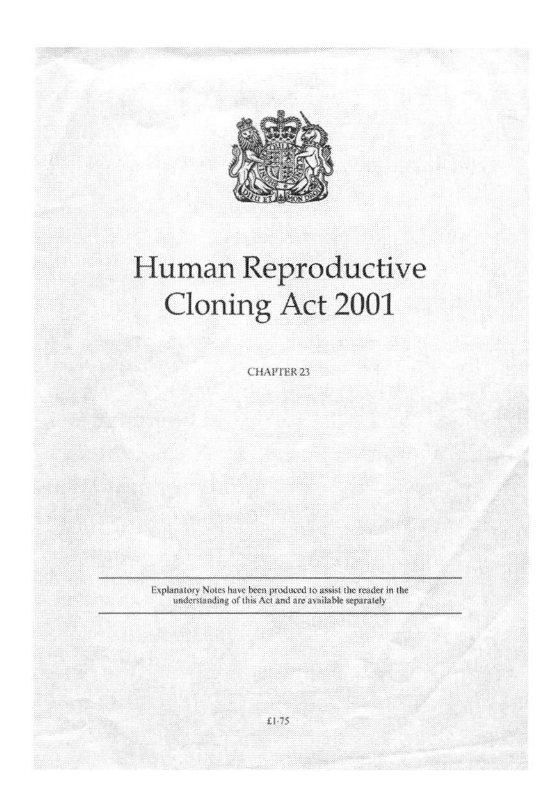
Human Reproductive Cloning Act 2001

CHAPTER 23

Explanatory Notes have been produced to assist the reader in the understanding of this Act and are available separately

£1·75

This 2001 legislation expanded the remit of permissible human embryo research in the UK to include purposes compatible with the use of somatic cell nuclear transfer to create "cloned" human embryos, for which a license was granted to Roslin in June 2003. *Courtesy of Her Majesty's Stationery Office.*

justifiable, if repetitive, engagement with something disturbing and wholly unresolved. Rather, it is to read "Dollymania" as a form of concerted, if inchoate, attention. After all, few animals have so succinctly embodied the complexity of human purposes and directions more obviously. But Dolly is an offspring of more than scientific ambition, just as scientific ambitions are never purebred offspring of science itself. Here again, genealogy does useful domestic service, possibly in a fittingly ovine manner, if we imagine it as something akin to herd instinct. Dolly came onto the scene for a whole flock of reasons that connect her, Roslin, bioscience, cloning, stem cells, and a myriad other biocultural entities together through lineages that are familiar, and even traditional, but newly hybridized, or mixed.

Like the sheep on which this book is based and from which it takes its primary inspiration, *Dolly Mixtures* deliberately resists the question of whether cloning humans is moral or immoral, instead taking Dolly's brief living presence at face value by attempting to describe the many logics, intentions, and hopes behind her making. It is this condensed past Dolly makes visible as a future intensification or direction. The position of *Dolly Mixtures* is thus that the answer to the morality of cloning Dolly lies in part in questions that were navigated prior to her emergence, and not only by scientists, in the dense historical accumulations that made her innovation desirable to begin with—both as a practical experiment and as a path to greater scientific understanding of basic biology. This is why our relationships to Dolly are neither trivial nor self-evident, either during her life or after her death. In the same way, Dolly's viability rewrote the biological rule book, canceling out major principles of biological development with a lively stamp of her newborn hoof, so this book attempts to begin to ask what kind of critical interpretive language might need to be devised or invented to begin to parse her cultural significance, arguing that such questions are not so much beyond as *before* the crude ethical dilemma Dolly is seen to pose about the limits and possibilities of the redesign of life itself.

At the level of basic social science, Dolly poses an age-old question about the social meaning of genealogy. Such formal questions, about classificatory kinship, animal totems, and the "biological facts of sexual reproduction"—themselves shaped by national, imperial, and colonial preoccupations (Povinelli 2002; Wolfe 1999)—motivated the emergence of disciplines such as anthropology and sociology in the early nineteenth century via questions about relationalities seen as purely biological at one level, but neither pure nor reducible to biology at another (Coward 1983; Franklin 1997b; Wolfe 1994b). The organization of paternity and maternity, blood ties and affinal connections, kinship and descent have long been recognized as highly variable and highly structured, both in humans and in other animals. Ties established through reproduction have long been regarded as uniquely significant to the production of distinctive kinds and types of relations, and thus to the wider ordering of social organization, while the precise structural qualities of such orderings—their formal character, that is—have been subject to extensive investigation and debate (Yanagisako and Delaney 1995). Remembering, therefore, that the reproduction of genealogy as a social form is inseparable from its biological dimensions, but that its biological dimensions have never determined its social form, and that this relation has long been a central concern of social science is an important antidote to the idea that cloning presents us with completely new kinds of questions (Strath-

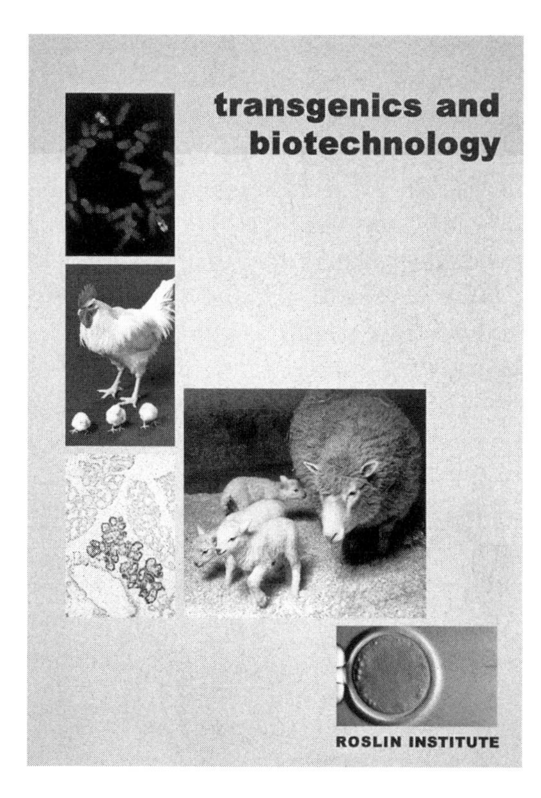

Both Dolly and Roslin are descended from a long and distinguished history of agricultural science in Britain which, increasingly in the post-war period, has become intertwined with biomedicine, including both drug development and assisted conception. *Courtesy of Roslin Institute.*

ern 1988, 1992a, 1992b, 1999). It does and it does not. Some of the questions cloning, biotechnology, and the new genetics pose have been the subject of intense discussion for over a century, albeit in a language that might at first sight appear more remote than this book argues need be the case.

In sum, *Dolly Mixtures* is a contribution to the ongoing reinvention of concepts and methods that can help define the issues at stake in the context of cloning, stem cells, and the new genetics. In an era in which theoretical work can easily be dismissed as lacking direct utility, this book argues that it is an essential resource for identifying both sources of knowledge that can be drawn upon to make better social sense of cloning, and areas where it is essential to learn more and develop new resources. As a preliminary, experimental, curious, and sheepish book, *Dolly Mixtures* attempts to open up new pathways of critical engagement with the future of genealogy and its pasts commensurate to the increasing technological power available to modify human reproduction and descent.

The risks in such a speculative venture are easy to identify, and *Dolly Mixtures*, like its willful namesake, pushes at the edges of scholarly etiquette in its leniency about borders and, some will protest exhibits uncivilized behavior. Adopting a discipline of inclusion risks compromising thoroughness, and a habit of attentiveness to the paradoxical and absurd dimensions of many of the sheep shifts related to cloning and the new genetics may dull the point of argument, or even make it disappear. Like sheep, this book is keener on exploratory foraging, endless rumination, and pushing over fence posts than it is on getting from the open pasture to the shearing shed so the wool can go to sale. If it is not easily penned in, I hope the advantages of rough grazing will be passed on in the benefit of its treadings, and outweigh the costs of persistent unruliness.

Chapter 1, "Sex," introduces the making of Dolly the sheep from the point of view of the biological facts of sexual reproduction, arguing that the project of making Dolly was a *remixing of sex*. This mix incorporated digital and virtual, as well as commercial and practical, dimensions through which genealogy, as well as sex, was reconstructed" and "reprogrammed." This argument is extended in chapter 2, "Capital," in which Dolly's value is investigated by returning to early English meanings of *stock* to explore the intertwined ideas of *trunk* or *stem* with *fund* and *tool*. Tracing these etymological and historical roots of the "narrowly English" meaning of stock-as-capital enable both the importance and the position of livestock, or cattle, in the coemergence of industrial agriculture and modern commercial markets to be more fully appreciated as continuous formal structures. These histories can be seen as both embodied and extended by Dolly, as well as by stem cells.

Chapter 3, "Nation," explores further why Dolly is a very British sheep, not only because she was born in Scotland but because of how Scotland became a "land of sheep" in an earlier era. This chapter thus also situates the uniquely stratified system of British sheep breeding in the context of other pastoral traditions that preceded it, such as that of the Romans, who first brought sheep to Britain, and in biblical traditions and literature, to which sheep-human symbolism is central. These themes lead naturally to chapter 4, "Colony," which explores the complex sheep exchanges linking Britain to her colonies, in particular Australia. As well as exploring the significance of sheep as white settler occupiers of colonial New South Wales, the return of Australian sheep to the mother country is explored in relation to IVF, cloning, and stem cells—areas of innovation which, as it turns out, are also "riding on the [Australian] sheep's back." Chapter 5, "Death," returns to Britain and to the scene of the 2001–2 foot-and-mouth crisis to continue sheep watching amid an epidemic imagined as the "ground zero" of biology "out of control." In this chapter, the question of how sheep death and sheep life are intertwined with the lifeways of rural communities is both a literal and a figurative one in that cloning, too, is an issue for which sheep are an "ethical frontier" on which the sheep-human interface is posed as a question, an opportunity, a threat, and a disquieting source of an anxiety it is difficult to come to grips with or resolve. The question of the sheep-human interface is taken up in the conclusion, "Breeds," as a means of reprising *Dolly Mixtures* and its central arguments about the intertwined genealogies of humans and sheep. If the question Dolly poses is that of "which way we are facing," and also "what we have to face up to," this can be reconfigured as a question of "frontier" confrontations, often represented through imagery of pioneering advances into the unknown. By taking a tip from sheep themselves, whose powers of facial recognition have proven far more enduring and complex than previously imagined, it is possible to consider whether the most important questions the Dolly technique asks us are less about where we are headed than who we already are.

Sex

It is not enough simply to say that sexual reproduction has become the dominant mode of propagation among organisms. One must go further. Cross-fertilisation, either continuous or occasional, is the really successful method of multiplication everywhere.
—Edward M. East and Donald F. Jones,
Inbreeding and Outbreeding

I began to envisage how to advance the project that I had been thrust into, while also satisfying my own desire for original research in developmental biology. Answer: don't just add DNA to one cell embryos. Add it to plates of cultured cells, and then make embryos from the cells that had taken up the DNA most effectively. In other words, as the 1980s wore on I began to see that the future of genetic engineering in animals lay through cloning.
—Ian Wilmut, Keith Campbell, and Colin Tudge,
The Second Creation

1

Clone is a term from botany, derived from the Greek *klon*, for twig. It refers primarily to regeneration via cultivation and propagation, in which reproduction is asexual and regeneration is of a part from a whole—as in the creation of a new plant from a cutting. *Clone* is both a noun and a verb, and has many figurative as well as practical usages. In all of these senses, cloning is synonymous with *copying*, its primary synonym. In this very general sense, it could be said that most reproduction occurs through a version of cloning to the extent that asexual reproduction, mitosis, or fission—creating multiple "copies" of an organism—is the most common, or standard, form of replication in living things—most of which are microorganisms and prokaryotes. A common distinction between reproduction and replication is the association of the former with sexual difference and the latter with more primitive copying abilities. However, a counterview would hold that microbes and bacteria, which have a famously

loose ability to exchange genes, are equally recombinant to the higher organisms with their much-vaunted abilities of sexual reproduction and the capacity to produce completely new individuals. Through cloning, even a single cell can produce a limitless population, and sterile lines can be reproduced, or grown, in perpetuity. The optimum all-terrain reproductive option of combining asexual division, or cloning, with the capacity to reproduce sexually exists among most species, including vertebrates such as birds and amphibians.

Sexual reproduction is associated with the higher species because it can be seen as more complex and advanced, whereas replication by mere division is associated with less developed life-forms.[1] The perceived advantage of sexual reproduction is that it combines replication with the production of genetic variation, or mix. Sex, in the sense of sexual reproduction, increases mix, which is seen to offer the advantage of maximizing adaptive capacity, or fitness, through variation. Mix, it could be said, adds flex—the ability to change.

In the account of making Dolly, *The Second Creation* (2000), Ian Wilmut argues that compared to asexual replication, sexual reproduction is expensive, dangerous, and inefficient: "Asexual reproduction is the obvious way to replicate: start with one individual, split down the middle, and then you have two, or many. By contrast sexual reproduction is a bizarre and even perverse way to replicate. Two protozoans that seek to reproduce sexually must first fuse, to produce one. . . . Sex, in short, is anti-replication. Replication implies that one individual divides to become two or more. But with sex, two combine to produce one" (Wilmut, Campbell, and Tudge 2000, 62). From this point of view, sexual and asexual reproduction are binary opposites: whereas asexual reproduction *amplifies one cell into two*, sexual reproduction *constricts two cells into one*. Together, the two processes offer opposite functions. Whereas asexual reproduction is efficient in terms of multiplication, sexual reproduction is a faster way to mix. The ability to combine sexual and asexual reproduction is consequently seen to offer the greatest advantage, and this appears to be the overriding pattern in the majority of species, in which the two mechanisms occur with roughly equal frequency and are often combined.[2] Significantly, this mixed approach to sex is exactly what Wilmut and his team sought to achieve technologically through the series of experiments that led to Dolly's birth.

Dolly's birth was also mixed in that she was designed to fulfill specific practical purposes, while being an experimental animal who would contribute to the understanding of basic questions in "pure" science. She was thus typically agricultural in that the project of her creation combined basic questions of genetics, or selective breeding, with commercial and industrial applications. As Wilmut writes of Dolly's significance, she is the animal model who confirms the viability of a technique that constitutes "the third player" completing a "trio of modern biotechnologies [that] taken together [take] humanity into a new age—one as significant, as time will tell, as our forebears' transition into the age of steam, of radio, or of nuclear power" (Wilmut, Campbell, and Tudge 2000, 18). Indeed, it would be cliché, but accurate, to describe Dolly as the proof of a new form of nuclear power—the power of nuclear transfer to advance the project of transgenesis that was Wilmut's overall aim.[3] As he summarizes, the value of the trio the Dolly technique completed, by becoming its means of application, lay in the facilitation of gene transfer.

> The point here is that the three technologies together—genetic engineering, genomics, and our method of cloning from cultured cells—are a very powerful combination indeed. Genetic engineering is the conceptual leader: transfer of genes from organisms to organism, and the creation of quite new genes, makes it possible in principle to build new organisms at will. Genomics provides the necessary data: knowledge of what genes to transfer—where to find them and what they do. Cloning of the kind that we have developed at Roslin and PPL [Pharmaceutical Proteins Limited] makes it possible in principle to apply all of the immense power of genetic engineering and genomics to animals. (Wilmut, Campbell, and Tudge 2000, 21)

This "powerful combination" of technologies is thus the metamix for transgenic sex, to which Dolly provided "the gilt on the gingerbread," according to Wilmut (Wilmut, Camble, and Tudge 2000, 20). She and the flock of Roslin sheep who were her contemporaries together confirmed the viability of a means of combining reproductive mechanisms that Wilmut argues "will take humanity into the age of biological control" (Wilmut, Campbell, and Tudge 2000, 24). This is the possibility "to grow animal cells in a dish, as if they were bacteria or cultured plant cells; and then transform these *en masse*; and then—

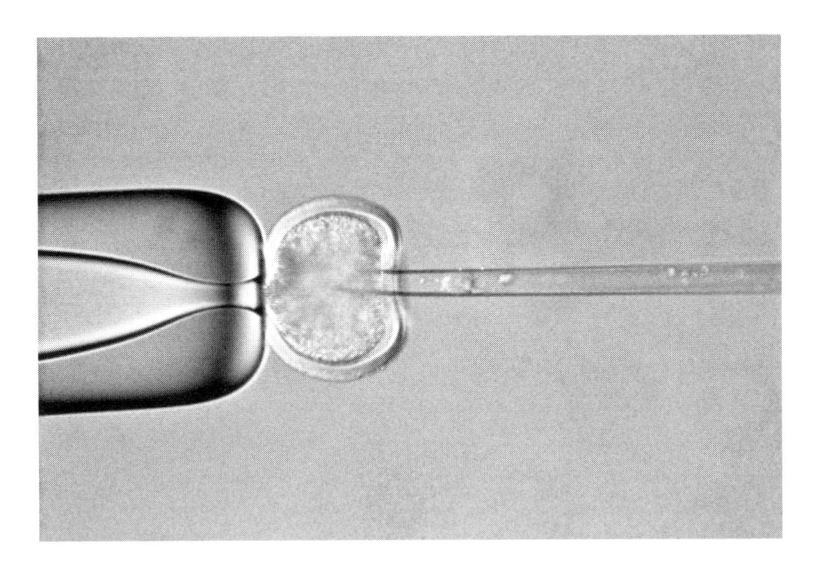

Microinjection imagery, such as this dawn of life's recreation in vitro, comprise a primary site of transformation in the meanings of sexual reproduction, which is here shown literally being constructed through the process of somatic cell nuclear transfer. *Courtesy of Roslin Institute.*

as is already carried out with bacteria and plants—grow whole new animals from the cells that had taken up the genes most efficiently" (Wilmut, Campbell, and Tudge 2000, 20).

This metamix of sex, in which the reproductive possibilities of plants, animals, and microorganisms are conjoined with biotechnological expertise in order to grow animals from cultured transgenic cells is what offers the totipotency Wilmut describes as bringing into being a new age—"the age of biological control."[4]

What is significant about sex after Dolly, then, is not that existing definitions and formations of it have been *transformed*, but that they have been, in a sense, sampled, *remixed*, resequenced, and provided with a novel means of amplification. Sex, which was never pure to begin with, is further hybridized through technological assistance to create a form of mixed-sex, known as the Dolly technique, somatic cell nuclear transfer (SCNT), or the second creation. Sex, in the sense of a reproductive mechanism, has been disassembled and *rearranged* through processes of reversal, recapacitation, switching, imitation, and transfer that allow it to be *redeployed* and *redirected*. The languages and practices of the remaking of sex that accompanied the creation of Dolly thus pose many questions for which sexual activity has become, as it were, a preliminary. That the questions her mixed-sex and multipurpose ancestry pose for normative conceptual regimes of sexual difference, biological reproduction, and animal kinds are inseparable from their ramifications in changing definitions of capital, nation, life, genealogy, and health is the main reason Dolly's mixed sex matters.

Dolly's sexual significance is also notable in the way her own ability to reproduce sexually confirms the scientific legitimacy of her vitality—by authorizing the scientific team that made her in the act of bearing her own viable offspring. In so doing, she models the technique used to create her and in a sense doubly—her own reproductive capacity confirming the viability of the technique by which she was created in a kind of variant on the progeny test (the progeny proof). As the model organism of the Dolly technique, her importance is at one level simply to embody it and to confirm its viability. Her body cells have been repeatedly analyzed to provide confirmation of her genetic provenance, and so Dolly *is* the technique that made her in some importantly literal senses. Technically, however, Dolly's complete viability was further confirmed by her ability to reach sexual maturity and breed naturally. This second reproductive proof confirmed both her ability to function sexually and the entirety of her sexual function—much in the same way as, but extending, the tradition in which only entire, or sexually complete, animals are allowed into official purebred lines.

The Difference Dolly Makes

Given how she was made, by merging two cells, in what sense is Dolly a clone? The word does not appear in the short letter to *Nature* announcing Dolly's birth, "Viable Offspring Derived from Fetal and Adult mammalian Cells" (Wilmut et al. 1997). Although Wilmut frequently refers to Dolly as a clone, he and his scientific partner Keith Campbell both acknowledge that the term is not strictly accurate. As Campbell states, "In the strict sense of the meaning, the animals produced by nuclear transfer are not true clones. Account must be taken of possible changes that occur in the genome during embryo and fetal development or while the cells are in culture. . . . Differences in the components of the egg cytoplasm would result in differences in the offspring. For example differences in the mitochondrial genome" (qtd. in Klotzko 2001, 10–11). In other words, *clone* is being used to describe Dolly in the absence of a more accurate term. She is not, technically, a clone in terms of a part regenerated from a larger whole, for her origins lie in two cells that were merged, or mixed, to make her.[5] She was also thus not descended from one parent, but from two. She was not created through "mere" division, and she is not even genetically identical to her "clonal mother" (the Roslin term for the sheep whose nuclear DNA was used to make her). Dolly is a genetic mixture—her nuclear DNA derived from a Finn Dorset ewe, her mitochondrial and other cellular DNA provided by a Scottish Blackface. Ironically, Dolly is a clone because she is different (from other sheep)—and indeed unique (among higher mammals). She is, as Campbell points out, also the product of her environmental influences, which in her case included being cultured, passaged, incubated, frozen, thawed, recultured, biopsied, electrified, and gestated by a series of surrogate sheep before she was even born.

Dolly is referred to as a clone because the technique used to make her belongs to the scientific history of experimentation in embryol-

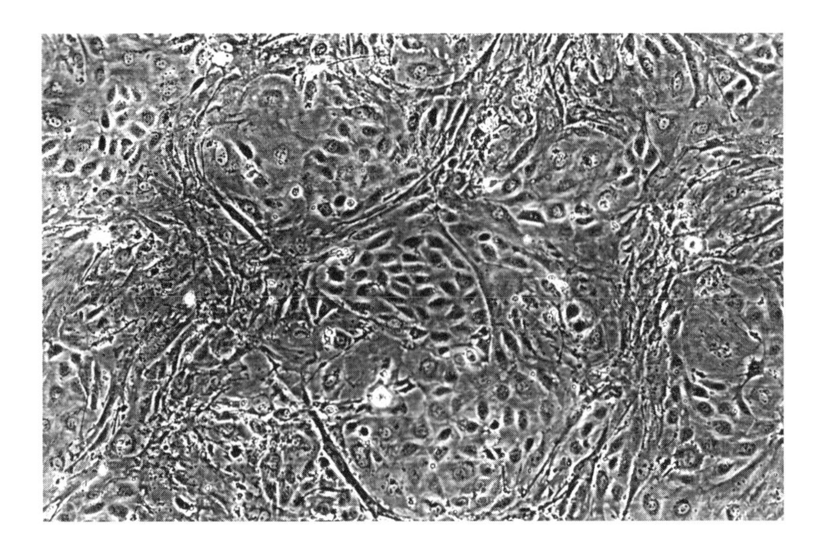

Dolly's lineage not only links her directly to cultured totipotent stem cells, such as the ones shown above, but also to a series of cloned ancestor sheep, such as Megan and Morag, who, despite being a different breed, were crucial to the emergence of Dolly, as well as later transgenic sheep such as Polly. *Courtesy of Roslin Institute.*

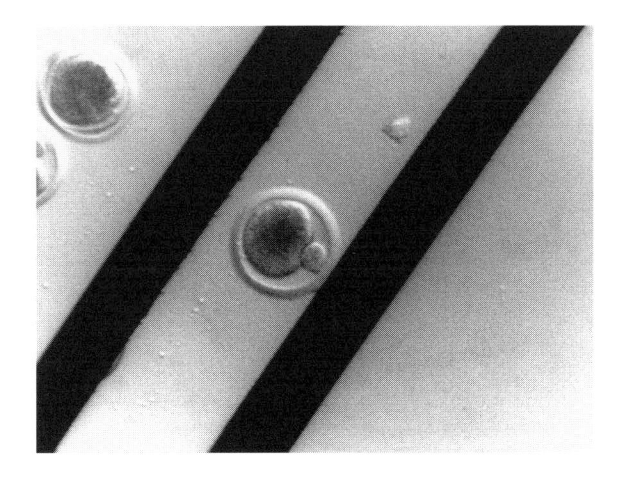

The reconstructed cell used to make Dolly was created through somatic cell nuclear transfer during the quiescent phase of cellular activity sometimes described as cellular sleep. An electric charge is used to complete this process. *Courtesy of Roslin Institute.*

ogy and genetics associated with cloning in a very general and imprecise way. *Clone* is the best way to describe Dolly more in terms of what she is not, rather than what she is: she is *not* an offspring of the usual method of mammalian reproduction, but of an unprecedented process of reproductive recombination. Hence, *cloned* signifies what sets Dolly apart, which, as we have seen, is both her path of ancestry, or origin, and her ability to embody this difference successfully, unlike all of the other attempted animals in her nonviable sibling cohort.

Before she was born, the cell line that would become Dolly, and with which she shares her nuclear genetic identity,[6] was cultured through several passages, meaning that the cells multiplied through division and were transferred, or replated, into petri dishes. This phase of Dolly's emergence is an essential component of the Dolly technique because it is during the period the cells are in culture that they can not only be multiplied but also modified through gene targeting. The fundamental key to the Dolly technique, then, is the ability to switch back and forth between sexual and asexual forms of reproduction.

The difference Dolly makes can thus be seen at several levels—which together express some of the paradoxical features of Dolly's life as a unique clone, a cloned singleton, and a sheep who was distinctive because she was "normal." Ironically, what the word *clone* means in relation to Dolly is that she was a unique individual because of the means by which she was created. Dolly's sex thus belongs less to a familiar economy of sexual difference but rather to a new scale of clonal difference. The difference cloning makes is not so much sexual as technical: it is a means to change sex in order to achieve specific technical goals, such as the more rapid amplification of flocks of transgenic sheep, for which somatic cell nuclear transfer, or remixed sex, is the most efficient mechanism.

Double Negatives

The technical sense of cloning that refers to the production of an identical organism (e.g., a vertebrate) from a parent organism has an unusually controversial history in modern biology, ironically intertwined with the same taint of fraud evident in uses of the term *clone* to refer to illegitimate copying (e.g., Gucci clones), transgressive sexuality (gay clones), and reverse engineering ("cloned" PCs). Something of a holy grail for twentieth-century developmental biology, claims to have successfully cloned animals have been subject to intense criticism and scrutiny in the past, and have even ended prominent scientific careers.[7] Long before the faked results of the South Korean scientist Woo Suk Hwang's cloning experiments were exposed, an almost implicit association of the science of cloning with fraud and deception already prevailed. David Rorvik's fictional account of cloning published in 1978, in which he so skillfully imitated scientific expertise as to fool even his publisher, famously tarred the science of cloning with the brush of scandal and deceit. The figurative senses of cloning as "imitation" and "simulation" are thus paradoxically exemplified by the history of the science of cloning—so much so that even the word *clone* sets off alarm bells of various kinds.

The dangerous illicit clone, its negativity doubled by both its figurative and historical associations, is generically and traditionally an abject embodiment of a particular kind of genealogical shame. Suspected of being a fake, a derivative, a copy, or a mere replicant, the clone is diminished by lack of a proper genealogy—and thus identity, substance, or origin. The pedigree of the clone is subaltern, in the sense of inferior and subordinate, because it lacks separation from the original, and thus a distinct identity. It is, rather, the identity *with*, in the sense of being identical to, its progenitor that makes the clone synonymous with a lack of the most fundamental kind—of individually defining substance. These pejorative associations with *clone* are combined and reproduced in its usages to refer to illegitimate sexuality based on narcissistic identification (gay clones) and slavery (either as "slavish imitation," or in the association of clones with a worker class of slaves or drones).[8]

Clones and cloning have figured prominently in the history of ideas, in the arts (literature, film, and theater), as well as in science. The genealogy of the clone can be traced through Greek mythology, Hebrew scripture, and the Bible, as well as being a staple figure in popular culture, advertising, and Hollywood cinema (Battaglia 2001; Stacey 2005). As we shall see, it is somewhat less surprising than it may at first have seemed that the world's most famous clone at the end of the twentieth century was a sheep. This benign domestic animal—a symbol of human innocence, Christian piety, and abject subordination—confronted the world in sheep-face, all but obscuring her threatening connotations as a clone. The Dolly grip into which the world's fascinated gaze was pulled up close and personal with a real, live clone in a hay-strewn Scottish paddock revealed her to be the epitome of harmlessness—a healthy, frolicsome lamb with an affable disposition.

A New Kind

Dolly's paradoxical uniqueness is one of many features that make her an animal that is "good to think with," as the anthropologist Claude Lévi-Strauss famously claimed about animal categories in general (1972). As well as being an intriguing mixture, she is a paradoxical multiple. Dolly is fascinating both because of what she literally is—a higher vertebrate cloned from an adult cell through a technique that was supposed to be biologically impossible—but also for what she represents culturally—an unfamiliar animal, an impossible animal, a counterintuitive animal kind. As Ian Wilmut suggests, "Dolly is the stuff of which myths are made. Her birth was *other-worldly*—literally a virgin birth; *or at least one that did not result directly from an act of sex*" (Wilmut, Campbell, and Tudge 2000, 233; emphasis added). For this reason, we can think of her as a Dolly morphism, a kind of mutational space in which cultural and biological categories, presumptions, and expectations are warped. It is difficult to get to grips with Dolly because she slips out of familiar kinds: her existence does not parse within familiar categories, or, rather, when she is fitted into them, they must twist and change to accommodate her unprecedented existence. Dolly is syntactically noncompliant within the normative arboreal grammars of reproduction and descent: her queer genealogy haunts the very basis of the formal biological categories that once affirmed the stability of a known sexual and reproductive order, following an unchanging pattern of bilateral, unilinear, and universal descent.[9] She is, in sum, an odd sort—not even a proper clone.

Dolly signifies both sameness and difference as a clone, but she also represents a fundamental confusion about how we tell them apart. Dolly is in this sense a totemic animal in the anthropological sense of being an animal entity who works to secure the very meaning of kind. Her kind is what Donna Haraway has defined as "trans." As Haraway notes, "The techniques of genetic engineering developed since the early 1970s are like the reactors and particle accelerators of nuclear physics: Their products are 'trans.' . . . Like the transuranic elements, transgenic creatures, which carry the genes from 'unrelated' organisms, simultaneously fit into well-established taxonomic and evolutionary discourses and also blast widely understood senses of natural limit. What was distant and unrelated becomes intimate" (1993, 56). Although Dolly is not herself transgenic in the sense of carrying genes from unrelated organisms, she is the viable offspring of a technique designed to make transgenesis more efficient, and she was designed precisely to "blast widely understood senses of natural limit." Indeed, like her trans-kin OncoMouse, Dolly belongs to a lineage in which, as we have seen, sex is used as a technique to transfer genetic traits nongenealogically (i.e., not via heredity). But what kind of kind does this make Dolly? What species of breed, or strain of "trans," is she, and how can we tell? What are her defining characteristics, and to what classificatory or kinship orders does she belong? What does her noncompliant existence do to the logics that stick these ordering systems in place? Does she confirm the order of things, or is she emblematic of their disorder?[10]

Perhaps Dolly belongs to a new order of animate, *trans-viable*, existence that is defined by being designed and made, or grown and built, rather than born and bred? Should her novelty occasion the emergence of a new Latinate denomination, *Ovis aries petri*? If Dolly's viability guarantees the functionality, and scientific validity, of a specific technique—somatic cell nuclear transfer—is she an animal who can only be properly classified in terms of how she was made? Since Dolly was made through a literal process of genetic drag, which enabled adult DNA to "pass" as newly youthful, is she uniquely biologically queer? Is this why Dolly's viability is haunted by the taint of vague and indeterminate insecurities that, even after her death, cannot be penned in? Do these insecurities stem from the fact that her origin story as a designer animal was so counterintuitive—or from the fact that building a sheep to order so is not news? Does Dolly's viability challenge the basic genealogical syntax on which so much normative social and biological categorization, or ordering, has been based? Or, in contrast, does her creation reveal how little that model ever really explained?

Double Trouble

As noted earlier, it is Dolly's simultaneous connotations of duplicity and singularity that are a likely cause of much of the ambivalence surrounding her creation. Fittingly, this fascination is expressed in the range of ways Dolly herself is literally reproduced as a cultural icon—her likeness(es) reprinted on everything from magazine covers to T-shirts, often in duplicate or multiple images. If Dolly stands for a change in reproductive form as well as function, and in reproductive code as well as substance, this is reflected in images that imitate the manner of Dolly's creation (copying) through the form of the image (as a multiple), calling to mind Andy Warhol's fascination with the view, aptly summarized by Hillel Schwartz, that "in our post-industrial age the copy is at once degenerate and regenerate" (1996, 257).

In all of these respects it is Dolly's connections to sex, genealogy, and reproduction that are distinctive, troubling, and significant. Her individuality, singularity, and uniqueness defy the fact she is a clone —a term associated with copying, multiplicity, and replication. The invisibility of Dolly's deviant biological origins, or queer ancestry, is compensated for in images of Dolly that double or repeat themselves.[11] This Warhol effect, by which Dolly's biological origins are represented by a "cloning" of her image, effects a substitution of visual reproduction for biological reproduction in a typical Dolly switch. Her associations with the power to create life, and with the deadly risks of this capacity, are paralleled in the way figurative images occur

In contrast to its high-tech associations, much of the equipment at Roslin is hand-made, such as the pipettes used by Bill Ritchie to conduct thousands of nuclear transfers on microscopic sheep eggs. *Courtesy of Roslin Institute.*

in a space of suspended animation, or "still life," at once imitating and replacing their objects in a substitute form. In sum, Dolly troubles ideas of reproductivity, of life and death, and of sexual difference and sameness by mixing them and thus altering not only their content and their meaning but also their function their order and their form.

The most obvious reason Dolly has attracted attention and comment is her much-celebrated and frequently debated embodiment of new forms of bioscientific potency, such as assisted conception and genetic engineering, that are rapidly reshaping the lives of humans, animals, plants, and microorganisms. Dolly represents the increasing ability to introduce new forms of transfer into the processes of reproduction and heredity, as well as the possibility of translating these into new applications that promise to improve agriculture and human health. In particular, she represents the possibility, or threat, of a kind of technology transfer—from sheep, *Ovis aries*, to humans, *Homo sapiens sapiens*. She thus stands for the desire to distinguish the animal from the human, and to prevent their mixture, while also, paradoxically, embodying their ever more proximate union—and the fallacy of such a dividing line between them.

In sum, Dolly embodies the long-standing debate over animal domestication—a topic about which there has been much renewed anthropological debate in the context of genetically engineered life-forms (Leach 2003; Cassidy and Mullin forthcoming). As I discuss further in chapters 3 and 5 (on nation and death, respectively), a key shift in debates over domestication has been the focus on whether domestication is a form of biological control (by humans over plants and animals), or the reverse. As Helen Leach argues,

> However it is defined, domestication was a process initiated by people who had not the slightest idea that its alliance with agriculture would change the face of their planet almost as drastically as an ice age, lead to nearly as many extinctions as an asteroid impact, revolutionize the lives of all subsequent human generations, and cause a demographic explosion in the elite group of organisms caught up in the process. Such unforeseen consequences are seldom discussed in the literature of domestication, perhaps because it is not in the nature of the species that started the process to admit that it isn't in control. (Leach forthcoming; references removed)

In this view, the genetically engineered animal is both a symptom of human overconfidence in biological control and the culmination of a lengthy process by which the drastic consequences of domestication have been unfolding.

The techniques used to make Dolly—and in particular their success, which she embodies—and thus introduce a new set of questions about domestication and "biological control," which is the role of reconstructed biology.[12] Of course she is not the first animal to be assembled in the laboratory, for there is a long history, particularly in mice, of "making" animals (Haraway 1997; Rader 2004). Neither is it only in the laboratory that animals are "built," as the purebred domestic dog is a product of deliberate reshaping through in-and-in breeding, engineering its constitution through selection (Haraway 2003). However, the successful piecing together of Dolly in the petri dish not only reconstructed the egg cell from which she originated but also some of the basic principles that structured modern biology pre-Her Ewe-niqueness.

This is why it is especially important to explore the implications of Dolly's cultured-up biology for the elementary formation of kinds and types, for example, by asking what happens to sex, breed, species, and reproduction when genealogy is retemporalized and respatialized. As a new kind of progeny, Dolly could be seen to belong to an emergent system of categorical difference brought into being by her existence outside the genealogical lines and linearities that formerly mapped the interior of the biologically possible. To put it another way, if biological categories are technologies for the production of logics of cultural difference, then how do some of Dolly's biological differences matter? And what do they offer as resources for thinking about kind and type more broadly, or the way these are organized along genealogical lines?

Biological Control

Writing about Dolly in the book that offers the most detailed account of her biological origins, Ian Wilmut claims that Dolly is "the most extraordinary creature ever born" (Wilmut, Campbell, and Tudge 2000, 15). He suggests that

> as decades and centuries pass, the science of cloning and the technologies that flow from it will affect all aspects of human life—the things that people can do, the way we live, and even, if we choose, the kinds of people we are. Those future technologies will offer our successors a degree of control over life's processes that will come effectively to seem absolute. Until the birth of Dolly scientists were apt to declare that this or that procedure would be "biologically impossible"—but now that expression seems to have lost all meaning. In the 21st century and beyond, human ambition will be bound only by the laws of physics, the rules of logic, and our descendants' own sense of right and wrong. Truly, Dolly has taken us into the age of biological control. (Wilmut, Campbell, and Tudge 2000, 17)

As we have seen earlier, part of what Wilmut means by biological control is exactly that—more precise abilities to design and create animals. Here, however, what Wilmut means by "the age of biological control" is that Dolly's existence has breached some of the biological limits formerly assumed to act as a kind of natural boundary, or constraint, on human activity (such as the idea that something is "biologically impossible," which, as he points out, has now "lost all meaning"). The age of biological control thus not only refers to a set of possibilities nonexistant before Dolly, and which radically alter the scale of the potential biological manipulation of life, but to the loss of the idea that there is anything like a biological barrier or limit beyond which humans cannot go. An implication of Wilmut's definition of control is that a shift has occurred from the idea of the biological as *subject to conditions*, which can be deciphered and understood, to a view of biology as *entirely unconditional*, and thus subject only to the limits imposed upon it from outside.

The ultimate condition of the biological pre-Dolly was the idea of the gene—a self-contained, self-replicating, immortal unit that recapitulates in its very structure the binary coding function it performs. The gene not only carried instructions but was a mechanism for copying them. *Genetic determinism* proved to be an unnecessary expression: the genetic was defined by its one-way instructional, coding or determining capacity. As Evelyn Fox Keller writes of the "dissolve" of life itself into a molecule, "The story of the double-helix is first and foremost the story of the displacement and replacement of the secret of life by a molecule. Gone in this representation of life are all the complex undeciphered cellular dynamics that maintain the cell as a living entity; 'Life Itself' has finally dissolved into the simplest mechanisms of a self-replicating molecule" (2000, 51). This view of the molecularization of biology would now appear best to characterize the period of postwar twentieth-century biology, culminating in the publication of the draft sequence of the human genome map in 2001. The new view of biology, the deconditionalized view of post-genomic biology is defined by a return to the cell—the first primary unit of the life sciences, overtaken mid-century by the gene, but back in the ascendancy in part because of Dolly (and vice versa, for she was, in a sense, an offspring of the cellular turn). New models of life as complex, autopoeitic, informatic, semiotic, and indeterminate now sit alongside the older models of an essentially bipartite division between genetic instructions and everything else. The new unconditional biologies of the age of biological control are primarily imagined as plastic, flexible, and partible. They no longer work to a logic of a fixed structural system, but to that of flexibly reengineered functionality. In fast-growing fields of post-genomic science, such as tissue engineering and computational biology, as in agriculture, the questions of what the biological *is* has become inextricable from what the biological *does* or can be made to do.[13]

As Ian Wilmut's descriptions of Dolly's creation make clear, her birth was one of the prominent late-twentieth-century events through which the power of genetic information was coupled into the harness of cellular reprogramming. The success of the U.S. scientists John Gearhart and James Thomson in deriving the first human embryonic cell and germ lines shortly after Dolly's birth further increased the attractiveness of genetic and cellular reprogramming (see Thomson et al. 1998; and Shamblott et al. 1998). In the first decade of the twenty-first century, this has proven to be a very lively arena in part because it is seen as the next logical step in the effort to understand and control gene function, now increasingly understood in terms of epigenetic effects. Post-human genome sequencing, this is the obvious missing link, which will need to be much better understood in order for the vastly increased knowledge about molecular genetics acquired since World War II to be translated into viable applications. This will require more complex understandings of how protein and biochemical signaling determine gene function in situ, and it is for this reason that the cell has become the new petri dish for studying the mechanisms regulating genetic expression. This broad shift to postgenomics, tissue engineering, and epigenetics has been occasioned by changes in how the gene is imaged and imagined, whereby the language of genetic codes, messages, and information has yielded to discourses of genetic pathways, switches, or constructs to be downloaded (and demoted) to become (but) one of several protein events that can be reversed, mimicked, or reengineered. It is these transformations—at once conceptual and technological, as well as commercial and political, and nationally specific—that the making of Dolly the sheep both demonstrates and performs.

In sum, one of the most important mixtures out of which Dolly is made is the blend of genomics and embryology enabling gene function to be mapped and redesigned in live cellular environments. This is what Ian Wilmut means by the age of biological control and is the reason why he compares its power to steam or radio—the tools of the industrial and communication revolutions. Wilmut's experiments were precisely designed to demonstrate what can be accomplished when techniques of cellular reconstruction are used in concert with cell culture and gene targeting to create an unprecedented level of control over biological functionality—indeed, one in which biology and its control proved identical.

As Wilmut explains, a more complete technical synthesis of biology and control is exactly what he was trying to achieve when he and

Between 1995 and 1996, Roslin scientists cloned three types of lambs. Taffy and Tweed, born in the same summer as Dolly, were cloned from Welsh Black fetal fibroblasts that were cultured and differentiated before being injected into enucleated Scottish Blackface eggs (cytoplasts). The two rams were named by Roslin scientists after Welsh and Scottish rivers accordingly. *Courtesy of Roslin Institute.*

Tracy was born in 1990 as part of early efforts to produce flocks of transgenic sheep who would express human proteins in their milk—or so-called pharm animals. The effort to improve on the process by which Tracy was created, and to more efficiently produce flocks of transgenic sheep like her, was the origin of the Dolly experiment. *Courtesy of Roslin Institute.*

his team made Dolly in 1996. He did not start out trying to clone a mammal, or to overturn the biological rule book, but to find a more efficient way to make a transgenic dairy animal that would carry and express a human gene—what we might call a "manimal." A dairy animal was used so that the system of lactation could be deployed as a switch to trigger gene action and provide a ready-made system for extracting valuable human proteins from the milk of high-yielding transgenic flocks. These proteins would then be used to make pharmaceutical products both for rare, untreatable genetic diseases such as cystic fibrosis, and for more common metabolic disorders such as diabetes and hemophilia.[14]

Dolly was one of a series of sheep created at Roslin on the way to the desired goal of viable cloned transgenic offspring, achieved in 1997 with a sheep named Polly. To chart Dolly's precision-engineered ovine lineage, Wilmut points to the elegant and much-admired experiments of the Danish agricultural biologist and sheep embryologist Steen Willadsen, who began cloning sheep by splitting two-cell embryos at Cambridge in the 1970s—more than a quarter of a century before Dolly was born. During the 1980s Willadsen continued to experiment with sheep, establishing and refining many of the techniques later used at Roslin to create Dolly. He was the first to use nuclear transfer to create viable (ovine) offspring in the mid-1980s, using nuclei from eight- to sixteen-cell embryos to make reconstructed cells that were transferred into the oviducts of surrogate ewes who carried them successfully to term.

At Roslin, the effort to clone sheep using nuclear transfer was initiated in order to establish a more efficient means of producing transgenic sheep that would speed up the process invented by Willadsen. In the older methods, transgenic sheep were made through the laborious process of injecting the desired gene, or gene construct, into early embryos, and then gestating them to term. This process had a high failure rate, but it resulted in the birth of Tracy in 1990. She could produce large quantities of desirable human enzymes in her milk, and she was eventually bred with a transgenic ram to create entire flocks of sheep like her.

The process of creating Tracy resembled the one used for Dolly in that it deployed a mixed reproductive strategy, hybridizing transgenesis and conventional crossbreeding. However, it remained time-consuming, inefficient, and expensive to create transgenic sheep in such a manner, and the original impetus for Wilmut's Dolly experiments was to find a more commercially viable means of transgenic animal production, as part of the larger goal of achieving, in Wilmut's words, "the genetic 'transformation' of animals and of isolated animal and human tissues and cells, for a myriad of purposes in medicine, agriculture, conservation and pure science" (2000, 17–18).

What this ability to transform animals, as well as "isolated animal and human tissues and cells," was seen to lack—in scientific, technological, and commercial terms—was a more flexible and efficient means of reproduction, and it was this retooling of transgenesis the Dolly technique secured. Rather than taking years of painstaking and largely unsuccessful efforts to create a single transgenic strain or variety, it seemed preferable to be able to create entire flocks in one season by making enough cultured transgenic cells to propagate as many animal starter cells as required. Such a technique would have the added advantage of enabling rapid adjustments to such a flock by using the cultured cell lines, or colonies, as test beds to refine and adjust specific genetic functions. In effect, by resituating the labor involved in transgenic breeding to the seedbed rather than the animal by creating bespoke colonies of immortalized cells that could serve as a kind of germplasm archive, it would be possible to more efficiently propagate desired animal traits. The transgenic stem, or stock, could be used both to seed new lines of offspring, and to provide additional root stock for related lines.

This strategy of transgenic breeding using cultured cell lines was already well established by the time Wilmut and his team at Roslin embarked on a series of sheep trials to improve them. As Wilmut notes, the problem was similar to one apparent in gardening: it began with the basic condition of the soil.

> It seemed as if the problem was simply to find the right conditions for culture which would allow [embryonic sheep] ICM cells to multiply without differentiating. Cell culture after all is a craft as much as it is a science; to a large extent improvements are made just by adding things and taking things away, and seeing what results. The behaviour of cells in culture can be modified by changing the conditions just as a gardener can influence the behaviour of his plants. (164)

In time-honored agricultural tradition, Wilmut's patient efforts at cell propagation were accompanied by a selective breeding program to test his results. Dolly's contemporaries—Cedric, Cecil, Cyril, and Tuppence (Poll Dorset rams)—were cloned from cultured embryo cells in the same breeding season as Dolly, as were Taffy and Tweed (Welsh Black rams), who were made from cultured fetal cells.

Dolly's immediate precursors, Megan and Morag (Welsh Mountain ewes), were cloned from nine-day-old embryonic cells, cultured through thirteen passages, and thus differentiated far beyond the point cloning by nuclear transfer was imagined to be possible. Together, the millennial Roslin sheep, made from different methods and different breeds, comprise the unusual trans-kin flock into which Dolly was born and with whom she lived as paddock mates for most of her life.[15] In turn, her successor, Polly, cloned from cultured transgenic cells and born in 1997, completed Wilmut's master plan that had begun, less than a decade previously, with Tracy.

Dolly is also the descendent of her parent or donor sheep, who are unnamed. The mammary cells used to make Dolly were derived from a cell line made from the udder of an elderly (six years old) and pregnant Finn Dorset ewe, which had been cryopreserved by PPL Therapeutics as samples to be used for research into ovine mammary epithelial cells. The cell lines used for the Dolly experiment were made of ovine mammary fibroblasts as well as epithelials. By inducing the cultured cells into a particular stage in the cell cycle (the GO, or "quiescent" phase), Wilmut and his collaborator Campbell, who specialized in cell cycle agendas, were able to "convince" twenty-nine of these cells to fuse with donor egg cells from Scottish Blackface sheep to make embryos, one of which became Dolly. Because the recipient (Scottish Blackface) egg cells were enucleated (i.e., had their nuclei removed), the transferred (Finn Dorset) mammary cells supplied 100 percent of the nuclear DNA for the resulting offspring. Biologically, this is what was supposed to have been impossible.[16]

The Roslin team's most influential achievement was thus to establish through a series of experiments, over almost a decade, that biological differentiation can be reversed, even in fully specialized adult cells. Dolly was supposed to be a biological impossibility because before her birth it was assumed that all cells "commit" as they develop, becoming particular kinds of cells, such as hair cells, skin cells, liver cells, bone cells, or heart cells. The assumption was that as cells differentiate and develop to become specialized cell types, they lose the capacity to become other kinds of tissue. Biologically, differentiation

Cedric, Cecil, Cyril, and Tuppence. These four genetically identical lambs were cloned from cultured embryonic cell lines derived from Poll-Dorset sheep, a lowland breed, completing the trio of adult, fetal, and embryonic SCNT experiments at Roslin between 1995 and 1996, all of which led to the birth of viable offspring. *Courtesy of Roslin Institute.*

Cloned from Welsh Mountain sheep, Megan and Morag were considered by Roslin scientists to be as great an achievement as Dolly, in that they confirmed the viability of SCNT using cultured cells that had been through several passages. Their birth in March of 1996 would have been more widely reported had it not coincided with Scotland's infamous Dunblane massacre in which 16 schoolchildren and their teacher were murdered. *Courtesy of Roslin Institute.*

Born in 1997, a year after Dolly, Polly and her sisters proved the viability of a more efficient method of producing transgenic sheep, by genetically modifying immortalized cell lines and using these as "seed beds" for new animal lines, thus avoiding the more costly and unsuccessful process of adding genes directly to embryos, used to make Tracy. *Courtesy of Roslin Institute.*

is defined as "a progressive developmental change to a more specialized form or function" (AHD). Wilmut defines differentiation similarly as "the process by which cells change in form and function as they develop and take on a specialized role" (2000, 342). Differentiation is thus a fundamental feature of biological development, and what was most important about the models of biological development and differentiation that preceded Dolly's birth was that their defining formal property was their directionality. Differentiation was seen to be progressive, linear, and irrevocable—as well as diminishing of cellular potency: by developing, cells specialized and changed, losing their other functionalities. This "biological fact" of development explained why a mammary cell could not become, say, a liver cell: it could neither travel backward to become newly totipotent,[17] nor could it then reverse direction again and go forward in a new trajectory of development. Differentiated cells were defined on a scale of irreversible temporality: they could not, as it were, revert to a prespecialized state in which their fate was open ended.

Ian Wilmut and Keith Campbell were among many scientists to demonstrate that even adult cells cultured into colonies were highly flexible, but theirs was the first team to create viable offspring out of a mammary cell that was both adult and had been through several passages in culture. This contradicted another basic principle of modern biology, namely, that adult cells (soma) cannot function as reproductive cells (germline). It turns out that they can. Wilmut named the process of recapacitating specialized cells he and his team devised "de-differentiation." Initially, he claimed the Dolly technique enabled specialized cells to de-differentiate in order to become other kinds of cells—and indeed even to behave like germ cells, with all of their primordial (totipotent) capacities recovered. Later, he suggested that de-differentiation was the wrong term because specialized cells *never differentiated in the first place*.[18]

Another way to describe what Wilmut originally termed de-differentiation is thus recapacitation. If specialized, adult body cells are induced to deliver functions they were formerly presumed to have lost, and if this recovery can be described as the recapacitation of cell functionality, then another way to describe what the Dolly technique enables is a retemporalization of biology. In other words, through biotechniques, the temporality of the biological is being rescaled, or even recreated. If specialized adult cells appeared to be going back in time when they were, in fact, always already in a germinal temporality and telos to begin with (i.e., because they never left it), then it appears that one of the basic formal properties of the biological was only "there" to the extent it was assumed it had to be.

The "discovery" of de-differentiation thus has fundamental consequences for the idea of biological rules or principles, the belief that these can be presumed as givens, and models of what they are, or are not. As Ron James of PPL Therapeutics describes the significance of the Dolly experiment: "Dolly is the first demonstration that the genome of an adult somatic cell can recontrol development following nuclear transfer" (Public lecture, London, February 13, 2001).What Wilmut describes as "biological control" James here defines as "recontrol," for which Dolly's existence is a "demonstration," establishing the principle "that there *is no loss or permanent inactivation of the genome during development*." Thus another way to approach the question of the post-Dolly biological would be to explore the relationship between biological manipulation and biological facts or principles. If biological facts or principles turn out to be reversible, is this reversibility now one of the most important principles of assisted biology? What if something as basic as cellular differentiation turns out to be so malleable it looks like it was never really an actual biological condition to begin with? If this disappearance of formal biological properties can happen through newfound capacities of biological control, then *what is the difference between the biology and the control*? Perhaps this is what de-differentiation more accurately describes.

In sum, Wilmut's Roslin team achieved cellular recapacitation by combining experimental control over the timing and staging of cell cycles and intercellular transfers with the harnessing of the egg cytoplasm in order to reprogram DNA. Exemplifying the shift described earlier away from an emphasis on gene function per se toward an emphasis on its situated action in a range of cellular environments, or gene effects, is Wilmut's stress on the powerful cytoplasm inside of the ovum which, as he put it, "tells the DNA what to do" (Wilmut, Campbell, and Tudge 2000, 142). The egg cell used to make Dolly was one hundred times larger than the mammary cell it was fused with, so that in sheer physical terms it overwhelmingly dominated the cellular environment of the two cells once they were joined together with a jolt of electricity, dissolving the cell wall of the mammary cell and reactivating its contents. Wilmut explicitly describes the egg cytoplasm as a supercomputer in another transbiological cross between virtual and genetic codes for conduct.[19] Reversing the usual determinism attributed to DNA as the blueprint or master plan for cellular development, the Dolly technique contributed to a new emphasis on *situated biological communication*, in which the powerful egg cytoplasm replaces DNA as the origin of developmental "instructions" (143).

Thus overturning in a blaze of microvoltage the masculinist legacy of the "passive" egg—so skillfully analyzed by Emily Martin (1991) in her account of how biological conception narratives recapitulate conventional gender stereotypes—Wilmut's super egg-oism relocates the "programmer" of early embryonic development in the egg's body plan, rather than in its nuclear DNA.[20] A key to this process that

Wilmut and Campbell describe as "crucial" to their success is the so-called "switch" from maternal to zygotic function that occurs when the DNA of the early embryo is "activated" (Wilmut, Campbell, and Tudge 2000, 118–23). It is the ability of the egg cytoplasm to initiate gene activation that allows its supercomputer-like reprogramming powers to reactivate older DNA, as if it were being turned on for the first time.[21]

The Conditional Biological

Looking back over Dolly's creation, then, it is possible to summarize several important legacies from her making, as well as from her meanings, in terms of how she stands as an iconic, or totemic, animal amid high-profile changes and developments within the life sciences. One of the most important of these, which will recur frequently in *Dolly Mixtures*, is the extent to which she results from a mixture of agricultural, commercial, industrial, and medical purposes. Similarly, her construction required a vast mélange of expertise—from large animal surgery and ovine IVF to "classical" embryology and microsurgical techniques to cell culture experience, as well as cytology and genetics. It also seems notable what a broad range of purposes her creation was intended to serve—from biomedical and agricultural to theoretical and zoological ones. The offspring of "pure science" crossed with venture capital, she is also the viable offspring of a hybrid partnership of public service and private industry. While the company that funded Dolly's creation, PPL Therapeutics, has been liquidated, the company that bought the exclusive license to the Dolly patents, Geron began attempting to make human cell lines with human embryos at Roslin in 2003. These shifts all form part of the field defined as transgenesis, which is rooted in a particular mix of genetics, cell biology, and embryology.

The legacies of the series of experiments that led to Dolly can also be measured in terms of changes in understandings of the biological, characterized by a shift away from the idea of fixed biological conditions and toward a "conditional biological" that is more flexible and plastic. A key component of this newfound plasticity of the biological is its reversibility, through which capacities presumed to have been lost can be reactivated. Hence genes are codes that program pathways of development, but they can also be reprogrammed. Differentiation and specialization, which were presumed to be unilinear and progressive, can now be turned back. Thus another key feature of post-Dolly biology is the extent to which genes have become *situational* and *contextual*, adding to their newfound flexibility, by demonstrating how they can be reinstructed.

If Dolly is an animal whose manner of coming into being suggests that biology is increasingly viewed in terms of a mixture of genetic and epigenetic effects, and in terms of cascades and pathways that take shape through situated interactions, switches, triggers, and constructs, then another part of the trans-biological mix Dolly demonstrates is the layering of organic, virtual, and digital understandings of vital processes such as reproduction. As we have seen in the narration of Dolly's conception from the Roslin scientists who made her, the idiom of computer programming provides the most satisfactory model of the role of the egg cytoplasm, while other aspects of her conception, such as the requickening of her reconstructed cellular origins in a shoebox-sized hand-made electrical apparatus effectively perform an imitation of biological conditions to produce a desired effect. In the mix of hand-made and high-tech lab equipment used to make Dolly is also suggested a blend of the manual, the microscopic, and the molecular that is epitomised by the primal scene of micro-injection, recapitulating the viewpoint of the micromanipulator, and providing a panoptics of regeneration via the digital screen image. Were we thus to describe Dolly's biology as a mix of digital, virtual, manual, and molecular technologies, we would be accurate in terms not only of how she was made, but of how her making is described, witnessed, and represented.

Above all, what is notable about Dolly's creation is the emphasis placed on the idea of biological control. This poses the most important question for which she stands, namely, the difference between more and less biological control—something, as chapter 5 of this book demonstrates, may be difficult to determine. A question we are left with is how to understand the forms of assistance to conception and genealogy Dolly makes possible, and what kind of "biological facts" result from these possibilities. To ask if this is where de-differentiation has some unexpected descriptive accuracy, I would like to close with one of the most common scientific images of Dolly and her tale.

Dolly's Body

In popular scientific representations of Dolly, we are often provided with a dual image—of Dolly herself, and a line diagram of how she was made. This pictorial sequence of stages is her tale trailing out behind her, depicting the techniques through which she was grown and made. Following this trail of scientific production from two cells, their merger, their transfer to surrogates, and their gestation to become Dolly, we move in and out of sheep's bodies, making sense of their novel interiors, while they are normalized by their perfectly ordinary external appearances. What are we to think of this shot-reverse shot safari by which we are shown the logic of Dolly's making?

The pairing of a photograph of Dolly with the increasingly legible and familiar image of micromanipulation recapitulates this inside and outside panoptic of popular scientific explanations of cloning by allowing us, as in *Jurassic Park* (dir. Steven Spielberg, 1993), to witness the miracle of life's creation in animal form—*and also how it was done* (Franklin 2000). In this sequence of images, we see both Dolly's high-tech microscopic origin and her unremarkable sheep's body, untroubled by the radical nature of her reconstruction under glass, or her biologically "impossible" interiority.

In the following chapters, this double take on Dolly is read both as a very old and familiar agricultural image—of the improvement of farm animals through technology and of the use of farm animals to facilitate progress and improvement—and something less easily legible—the unprecedented biological totipotency she embodies. In turn, we can try to answer the question of what Dolly does to sex within the wider frame of how she belongs to lineages of capital, nation, empire, and science, as well as how she troubles these legacies.

2 Capital

The appearance of capital as an independent and leading force in agriculture does not take place all at once and generally, but gradually and in particular lines of production. It encompasses at first, not agriculture proper, but such branches of production as cattle breeding, especially sheep-raising, whose principal product, wool, offers at the early stages a constant excess of market-price over price of production during the rise of industry, and this does not level out until later. Thus [was the case] in England during the 16th century.
—Karl Marx, *Capital*

I am very aware of how much Keith and I have owed to venture capital.
—Ian Wilmut in Wilmut, Keith Campbell, and Colin Tudge, *The Second Creation*

A stem cell, as its name implies, is a cell that can "branch out" like stems of a tree and form more than one cell type. At minimum, to be a stem cell, as opposed to all of the cell types that inhabit most of the tissues in our bodies, a cell needs to be able to divide into two cells, one of which will be another stem cell similar to the original and one of which will change, or "differentiate," into another cell type.
—Michael D. West, *The Immortal Cell*

Like her queer connections to sex and modes of reproduction, Dolly's relationship to capital can be described both in terms of how she has extended its existing meanings and how she has transformed these through excess. Like her genetic identity, Dolly's economic value and her unique significance for emergent biotechnological economies lay in her multi-functionality resulting from the fact that she was made and grown as a cell before being born and bred as a sheep. Above all, Dolly was valuable because she was viable—a viable offspring and an animal model for a technique that confirmed a new means of propa-

gation from cultured cells. She was thus a "capital" animal both in the sense of being a principal animal, or base, for a new line or branch of production (of both animals and products), and in the sense of being the initial, original, or primary animal inaugurating an animal lineage that could perform the function of being a capital stock, or fund, out of which future profit could be generated. As this chapter argues, Dolly manifested capital's old and new, binding the oldest definitions of capital as "stock" into what Charis Thompson (2005) describes as the "promissory capital" of bio-futures markets in health, regenerative medicine, and stem cell manufacture.[1] Indeed, Dolly's value was precisely to be found in her ability to *unite* these disparate systems into a new form of totipotency. She thus both epitomized and vastly extended the meaning of livestock, or breed wealth,[2] through her embodiment of new forms of biological control over an expanded array of reproductive functions.

Significantly, although she was purposefully conceived and designed as an animal model of new forms of capitalized reproduction, Dolly was not a commercial animal in the sense of belonging to a particular market sector or product line. She was not herself a trade commodity, like other sheep, and, indeed, there are as yet no successful commercial markets deriving from her creation.[3] As this chapter argues, Dolly is neither fixed, circulating, nor floating capital, but, rather, protocapital, meaning that she is a kind of capital primordium, or source. She formed part of a series of experiments that combined commercial, agricultural, and purely scientific purposes in the effort to solve a specific practical problem. Her biology was not itself marketable, but provided a conduit, pathway, or line to future capitalization, and her viability secured a patent portfolio for a technique that may prove instrumental to biocapital's economies of vitality.[4] However, Dolly's immediate economic value was realized in the merger between Roslin's biotech company, Bio-Med, and the California biotech company Geron to create Geron Bio-Med in 1999, whose research priorities lie in stem cell research and regenerative medicine (Franklin 2001b, 2003b). Both Dolly's venture-capital value and the scientific value of the Dolly technique, then, derive from what might be called the stem technologies of cloning.

The unpredictability of stem, or proto-, capital is evident in the rapid transformation of the Dolly technique from a specific method of sheep breeding into a general platform for many kinds of cellu-

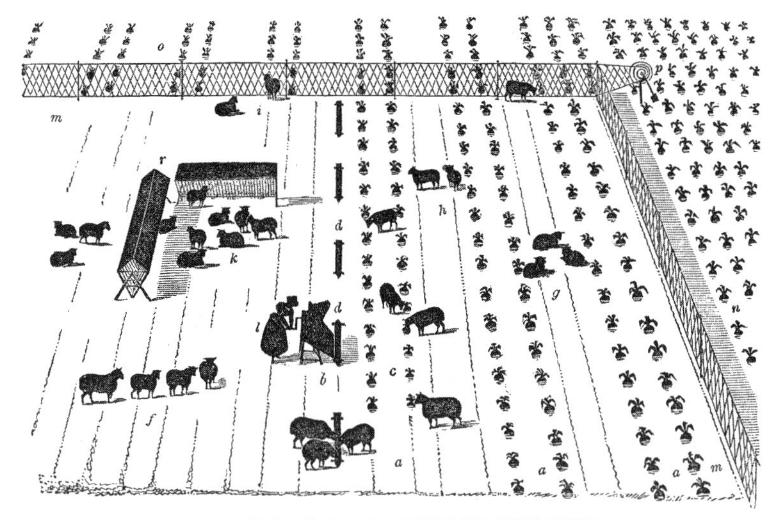

THE MODE OF OCCUPYING TURNIP-LAND WITH SHEEP.

The introduction of the cultivation of turnips, clover, and artificial grasses proved a turning point in English agricultural progress in the eighteenth century. Farmers could carry more and larger stock, which in turn improved the soil with their manure, creating more arable land. Sheep in this early context, then, could also be seen as "bioreactors." *Reprinted from Henry Stephens,* The Book of the Farm, *vol. 1, 2nd ed. (Edinburgh: William Blackwood and Sons, 1851).*

lar reconstruction, cultivation, and propagation, which increasingly involve human cells. As we have seen, Dolly's birth resulted from an experimental program dedicated to the production of transgenic dairy animals in order to extract valuable human enzymes in their milk. These experiments formed part of a carefully organized scientific and corporate plan to create a product and a market for the enzyme Alpha 1 Anti-trypsin (AAT), used in the treatment of cystic fibrosis and emphysema. As noted in the previous chapter, all of the Roslin sheep were steps along the path to the production of Polly, the transgenic bioreactor sheep, born a year after Dolly in 1997 and embodying the denouement of the joint Roslin-PPL venture to produce a founder animal for herds of sheep which could be multiplied, propagated, and improved in the petri dish.

As we have also seen, Dolly was not even the most important step in the path to Polly, as her birth could be interpreted merely to have reconfirmed the principle established by her precursors Megan and Morag, and even earlier Tracy, and her success as an experimental animal did little to further the ends she was originally designed to serve. As is now well known, somatic cell nuclear transfer turns out to be a very inefficient method of breeding, with high levels of pathology and morbidity, as well as inconsistency. It is unreliable, expensive, and, some say, unethical. It turns out it is not even necessary, as transgenic animals can be bred conventionally to fix the inheritance of the introduced trait. Nonetheless, the Dolly technique has become a literal hotbed of bioscientific innovation and one of the most important meeting points between practical innovation in agriculture and general biological interest.

Born of a merger between practical and hypothetical questions, Dolly is what the Danish scientist Steen Willadsen calls the "supreme example" of a sheep produced for commercial reasons but of much greater "general interest" to biology (qtd. in Klotzko 2001, 45). Initially envisaged as the "royal route" to increasing the efficiency of transgenesis (Wilmut, Campbell, and Tudge 2000, 54), Dolly's viability soon became the guarantee for a technique with much wider implications. The Dolly technique was seen not only to revolutionize understandings of biological processes but to open up new pathways for cell reprogramming and recapacitation. These in turn were seen to lead to the possibility of new forms of health care (regenerative medicine) and new branches of biomanufacture (stem cells) and tissue engineering.

This chapter explores Dolly's value as a form of livestock in order to emphasize what she provides as a kind of principal or base. In the same way she is multiply other things, she is multiply capital —a capitalicity. As such, Dolly helps us identify some of the emergent mechanisms and resources of biocapital on which future biocommerce is envisaged and imagined (and may or may not be based). She prompts us to ask: In what forms is the capitalization of vitality rendered viable? And: What are the techniques that allow vitality to be capitalized?

I begin by examining the relationship between capital, stock, and livestock, before turning to specific examples of stem cells, which are, in a sense, the most significant offspring of the Dolly experiment.

These examples build on the description of biology as virtual and digital in the previous chapter to explore the forms and techniques through which vitality is being capitalized via recapacitation, while also revealing some of the important differences between biocapital and biocommerce, as well as their weak connections. In turn, this provides the opportunity to rethink capital not only in terms of what it is or does but in terms of how it is being reconstituted in contemporary contexts of knowledge production, health, science, and medicine.

Taking Stock

Across the six definitions of *stock* to be found in the *Oxford English Dictionary* are more than two hundred separate definitions, running to thirty-four pages of examples. The earliest and primary sense of *stock* is of a trunk or stem, as indicated by its most persistent usages—in passages from the ninth and tenth centuries through to the nineteenth and continuing into the present—to refer to a trunk as a solid base, or generative stem, as in root stock, stock still, or lock, stock, and barrel. Combined in these early senses is the botanical idiom of roots, or movement, with the image of a stump that stands still, and could even be stone. It is the idea of stock as a solid block of wood, or base, which leads to its identification with possessions, tools, and the basis for trade. These are combined with the idea of *stock* as the origin of lines, or lineages, that diverge or spread. Varying combinations of these early meanings are invoked in the earliest figurative uses of *stock*, dating from the fourteenth century onward.

Definitions of *stock* include:

- The source of a line of descent; the progenitor of a family or race;
- The first purchaser of an estate of inheritance;
- The original from which something is derived;
- A line of descent; the descendants of a common ancestor, a family, kindred;
- A race, ethnical kindred; also, a race or family (of animals or plants); a related group "family" (of languages);
- An ancestral type from which various races, species, etc., have diverged;
- Pedigree, genealogy: a genealogical tree;
- Kind, sort;

- Feudalism: a serf by inheritance;
- Inherited constitution, breed.

Sixty-five subsequent senses of *stock* are listed under its primary entry (as a noun). These range from simple objects such as boxes, mallets, troughs, and handles to specialized implements ("a roller for a map") and various supporting technologies ("a stand or frame supporting a spinning wheel or churn"; "the support of the block in which the anvil is fixed, or of the anvil itself"). Included also are numerous references to tools and parts of tools, instruments, or weapons, as in "the heavy cross-bar (originally wooden) of an anchor"; "the hub of a wheel"; "the wooden portion of a musket or fowling piece."

The sixth, succinct, sense of stock, "a fund, store" (which is, as we shall see, in fact two senses), is noted by the editors of the *OED* to contain definitions that are both "obscure" and "blended." The very definition of *stock* is thus a blend of "different lines":

> The senses grouped under this head are not found in any other Teut[onic] languages except by adoption from English. Their origin is obscure, and possibly several different lines of development may have blended. Thus the application of the word to a trader's capital may partly involve the notion of a trunk or stem (branch I) from which the gains are an outgrowth, and partly that of "fixed basis" or "foundation" (branch II): cf. fund. . . . The application to cattle is primarily a specific use of the sense "store," but the notion of "race" or "breed" (branch III) has had some share in its development.

Definitions forty-seven through fifty-three have the sense of "fund" and are largely monetary or financial, as in "a sum of money set apart to provide certain expenses"; "the aggregate wealth of a nation"; "money, or a sum of money, invested by a person in a partnership or commercial company." It is in this subsection, related to the idea of a principal fund, that several uses of *stock* refer to capital. In these definitions, stock is again both fixed (as a sum) and fluctuating (in its value). Following are a few examples:

> 48a. A capital sum to trade with or to invest; capital as distinguished from revenue, or principal as distinguished from interest.
>
> . . .
>
> 50a. The business capital of a trading firm or company. *in stock* (said of a person): in the position of a partner.

. . .

52a. The subscribed capital of a trading company, or the public debt of a nation, municipal corporation, or the like, regarded as transferable property held by the subscribers or creditors, and subject to fluctuations in market value.

It is from these latter definitions that the modern meaning of *stocks and shares* derives, through what the *OED* editors describe as a specifically British "narrowing" of the meaning of *stock* to refer to "shares" defined as "capital of a public company . . . *when it is divided into portions of a uniform amount*."

The second half of the sixth sense, "store," is evident in definition 53a: "A collective term for the implements (dead stock) and the animals (live stock) employed in the working of a farm, an industrial establishment, etc." This definition both blends the two meanings of sense 6—"fund" and "store"—and suggests aspects of earlier definitions, in particular the strong association between *stock* and the uses and parts of tools or containers. The explicit opposition between "dead" tools and "live" animals strengthens this association, since the term *livestock* implies that such animals are instruments or tools. This definition also suggests that the coupling of live and dead stock together in this way, as a collective term, is characteristic of agricultural or industrial, and of English or British, contexts.

The next definition (54a), of livestock—defined as "the animals on a farm; also a collective term for horses, cattle, and sheep bred for use or profit"—similarly combines "fund" and "store," to which we can now add "tool" and "movable," as in an earlier definition (49c) of *stock* as "movable property." It is for this reason that livestock is followed in definition 54b of *stock* as "applied to slaves," followed by two nineteenth-century American quotations—from *Webster's Dictionary* (1828–32) and the feminist abolitionist Harriet Martineau (1837), referring to the slave trade in the Caribbean and the Southern U.S. states.[5]

Considering the earlier characterization of these definitions as exclusively British, obscure, and blended, *livestock* itself turns out to be a hybrid mixture with a distinctive pedigree. In the term are combined the meanings of *stock* as tool (tool part), and fund (or capital), with, as is also pointed out above, the earlier senses of race or breed. What also proves notable is the opposition, particularly relevant to the conversion of capital into shares, between a foundation or base that is fixed, solid, and stumplike (or "solid as a rock"), and one that is movable, divisible, partible, and in other ways flexible and capable of expansion.

The key component of the definition of *livestock* (54a) as farm animals bred for use *or profit* is the introduction of the use of such animals to generate surplus income—a definition that harks back to the long-standing equation of *cattle* with *capital*, but with a more specifically articulated connection to capital*ism*, in the use of the term *profit*. What is notable about the way the etymology of *stock* can be read as a historical narrative combining the emergence of distinctive forms of genealogy, property, and economy (now so unremarkable as to be obvious) is that this history is so well trodden by sheep. That this historical transformation of animals into stock should lead so directly to the propagation of human stem cells gives reason for pause, providing the question of alienated reproducibility with another modern, and yet archetypal, turn.

The Stock Market

The relevance of these definitions to Dolly as a capital animal is, of course, substantial, and indicative of how she unites very old and very new uses of the term.[6] Stem cell propagation, the new field of scientific endeavor and biotechnological investment engendered in part by Dolly's viability, are equally intriguingly overdescribed by the etymology of *stock*—most obviously in its primary sense of "trunk" or "stem," but also in terms of the meanings of "fund," "tool," "instrument," and "source." Indeed, stem cells unite the meanings of *stock* as form, lineage, kind, original, and type with its other meanings of genealogy, derivation, descent, race, family, constitution, and breed to the extent that we might think of stem cells as "life stock," in much the same way they are promoted as forms of life insurance, or in the language of banks and banking as the basis for venture capital speculation.[7]

Like domestic animals, stem cells are the product of historical accumulation and careful husbandry. For the same reasons the industrialization of animal breeding first took place in Britain—and Britain became the stud stock capital of the world in the nineteenth century, largely through sending its sheep to its colonies, so too is it regarded as the stem cell capital of the global biotechnology industry. As we shall see in the next two chapters, Dolly is a very fitting compliment to her British heritage, as are her primary "descendants" in the form of the stem cell lineages created the same way she was, and thus of her same "kind" in both the old sense of "species" and the new sense of "good breeding."

As Marx noted of sheep in *The German Ideology*, they are "*malgre eux*, products of an historical process" (1965a). This historical process, as he notes in *Capital*, manifests itself "gradually" through "particular lines of production," as, for example, in sixteenth-century England during "the rise of industry" through the sheep and wool trade. To quote again from the epigraph to this chapter: "The appearance of capital as an independent and leading force in agriculture does not take place all at once and generally, but gradually and in particular lines of production. It encompasses at first, not agriculture proper, but such branches of production as cattle breeding, especially sheep-raising" (1974). The relationship of the sheep and wool industry to the "rise of industry" is thus crucial to the emergence of capital as an independent and leading force in agriculture. It is the mercantile activity and enterprise associated with a principal pastoral product, wool, that enables a fund or store of profit to facilitate the emergence of "agriculture proper," or what might be called capitalized agriculture. It is in turn the industrialization of "branches of [agricultural] production" ("especially sheep-raising") that establishes the crucial infrastructure for the emergence not only of industrial capitalism but the displaced proletariat, the colonial settler, and the expansion of the British Empire as we will observe in subsequent chapters, once again following the paths of sheep.

As Donna Haraway has so eloquently demonstrated, transgenic animals constitute not just mixtures of animal, human, and machine but dense "implosions of natureculture" born of the myriad conceptual couplings that are "world shaping" in their consequences—such as "military-industrial," "bio-informatic," or "genetically-engineered." It is these vast and ordinary imploded forces of quotidian existence that are figured by Haraway's famous cyborg, whose "ontology . . . gives us our politics" (1997, 150). Haraway's cyborg is both an embodied organism made up of fiction and flesh and an analytical tool—an analogy, a figuration, a digging stick, or a symptom. Sheep perfectly embody this cyborgian ontology by which what they "are" cannot be extricated from what they have been made and bred to be

The lever turnip-slicer for Sheep. This piece of equipment, itself a form of stock, is typical of the new machines devised and perfected during one of the most famous eras of industrial progress in British agriculture when it could be said that the change of scale later analysed by Marx in the ability to cultivate arable land was made possible in no small part through engineering. *Reprinted from Henry Stephens,* The Book of the Farm, *vol. 1, 2nd ed. (Edinburgh: William Blackwood and Sons, 1851).*

The association of sheep with wealth and prosperity is widespread as indicated by this Chinese sheep bank, which is fleeced in coins and thus represents the value of sheep not only as stock, capital, or even biocapital—but money. *Photograph by Sarah Franklin.*

in accordance with the same principle fictions by which Haraway argues her readers are also shaped.[8] Hence, as she argues in her more recent manifesto about companion species, based on the domestic dog, "flesh and signifier, bodies and words, stories and worlds" are combined in the "remodeling and remolding" out of which human and animal bodies take shape (2003, 20). Animal domestication is a model of this process, as are its effects on humans. As Haraway notes, "the domestic animal is the epoch-changing tool, realizing human intention in the flesh" (2003, 27–28), but with equal consequences for the domesticators. "It is a mistake," Haraway writes in her account of canine domestication, "to see the alterations of dog's bodies and minds as biological and the changes in human bodies and lives, for example in the emergence of herding or agricultural societies, as cultural" (2003, 31).

Haraway's point of view, turning the lens of domestication back onto the human via the dog, works as well with sheep, who have had many shaping effects on human societies. Indeed, Haraway's points are to a degree anticipated by sheep historians, who note that pastoralism rests on an "ecological basis of domestication" that is behavioral, psychological, and sociological, as well as biological (Ryder 1983, 3). As the prolific sheep and wool historian M. L. Ryder observes in his magnum opus, the succinctly titled *Sheep and Man*,[9] "The domestication process can no longer be given an economic or religious explanation, but must be considered as an extension of the ecological relationship between sheep and man that was brought about by exploiting behavioural characteristics . . . . Why sheep were domesticated becomes a less important question than how when the ecological basis of domestication is understood. The key biological process appears to have been the 'imprinting' on a young animal of a human being in place of its mother" (1983, 3). Dolly extends the mutual imprinting of sheep and humans that has occurred for millennia and deepens the meaning of domestication to comprise their mutually molecular interiors, now being mapped as genes and "trained" as cell lines. From the perspective of domestic capital, this intimate extension of the pastoral economy—and, indeed, of pastoral care—has many important precedents. From this perspective, Dolly poses a question both as a domestic animal and as the embodiment of a technique with the potential value to colonize or domesticate human cells. As cellular behavior is "tamed" and the reproductive powers of cells are more directly controlled, the latter, too, will become purebred lines used for crossing and mixing, as well as for particular kinds of work.[10] Like good working dogs, their agility will be at a premium, and their obedience to human commands will depend upon highly disciplined attention from their handlers.[11]

Pastoral Capital

Clearly, sheep and pastoralism have been a major moving force in history, not only in Britain in the seventeenth century but for most of Europe, Africa, Asia, and the Middle East.[12] From preclassical societies such as the Phoenicians and Carthaginians, who may have raised the ancestors of modern Merino sheep, to Ancient Greece and Rome, where fine-wooled sheep were bred to clothe the highest social ranks and the emperor, to the earliest monopoly in Europe—the Spanish Mesta overseeing the transhumances—sheep and wool have been at the base of human economy and sociality. The remains of semi-domesticated sheep were found in the earliest Neolithic settlements in northern Europe, and they are the animal most often mentioned in the Bible, which draws endlessly on pastoral analogies. As Marx noted, "Were the term capital to be applicable to classical antiquity—though the word does not actually occur among the ancients . . . —then the nomadic hordes with their flocks on the steppes of Central Asia would be the greatest capitalists, for the original meaning of the word capital is cattle" (1965b, 79). Hence not only are sheep "good to think with" in terms of the history of capital but, as Haraway suggests, they have been integral to its changing shape. This is newly evident in the way Dolly and the Roslin sheep have passed their genealogies into the remaking of human health and the establishment of a bank of human embryonic cell lines, which has the potential to transform understandings of vitality as a (re)productive force.

Stem cells, then, like sheep, provide powerful models of the ways in which capital in the older sense of *stock* derives out of a combination of genealogy, property, and instrumentality. Stem cell science centrally depends on the primordial potency of cells and cell parts to provide its principal source, foundation, or "trunk." As the basic unit of all living things, the cell and cellularity constitute the fundamental elements of vitality, and their experimental study and observation is at the root, base, or origin, of modern biological science. Stem cells are special because of their multiply regenerative, or totipotent, capacities and the fact that these abilities can now be harnessed, reprogrammed, and redirected. The tools to achieve these ends are derived from an entire gamut of industrial, agricultural, clinical, and scientific techniques and applications. In particular, it is the ability to *engineer* new lines of descent from human embryonic stem cells that is envisaged as a new origin, or "dawn," for the biosciences and biomedicine. However, the lines of application derived from this new fund of biopotency are yet to be drawn as the science itself remains similarly embryonic in its development.

Primordial Flex

Stem cells feature prominently in both popular and scientific accounts of life as a flexible assemblage of components, capable of being reorganized while preserving its core vitality, in what some commentators have described as a new era of plastic biology.[13] The following opening paragraph from an article describing stem cell technology from the European Commission's research and development newsletter, *RTD Info*, offers a typical explanation of the features seen to make stem cells special.

> At birth, human beings are made up of approximately 100,000 billion cells belonging to 200 different categories (nerve, muscle, secretory, sense cells, etc.). Each of these groups is able to effect a number of very specialized tasks. As the body develops, the cells multiply by a process of division: when tissues deteriorate or wear out, it is generally the cells in the vicinity of the damaged zone which proliferate and try to compensate for the losses. Over time, however, this regenerative ability is progressively lost and ultimately disappears in many vital organs. (European Commission 2001, 4)

In this description, cells are classified in terms of quantity (100,000 billion cells at birth), type (two hundred different categories), and function (or "effect"). Cellular function is described in terms of "a number of very specialized tasks" including multiplication, division, replacement, specialization, proliferation, and regeneration. These are the key components of cellular effectivity, which are in turn organized economically, in terms of production and loss. Vitality is the outcome of the successful replacement of cells, and age, or diminished vitality, results from the waning of this capacity.

> This is why the discovery of the role and properties of stem cells (known as *multipotent* when they can form several types of cells and *pluripotent* when they can form all of them) brings new and exciting prospects. Tissues formed from cells so specialized that they are virtually unable to be renewed could—if damaged—be "reconstructed" through the addition of a sufficient number of stem cells. In any event, that is the underlying idea of what is hoped is a new field of medicine in the making: regenerative medicine. (European Commission 2001, 4)

Stem cells are important *because* they are exceptional. They are "the exceptional exception" precisely because they offer unique regenerative capacities: "Stem cells are a double exception to the rule of cell specialisation—hence their interest. Not only are they able to reproduce identically (and exceptionally quickly) throughout their lives but, more importantly, they are able to differentiate to form several (sometimes in very large numbers) distinct cell types" (European Commission 2001, 4). Stem cells, then, generate interest because they are multitalented multipliers. "Not only do they reproduce identically," but they "are able to differentiate." In this account of stem cells, they are doubly valuable because they are a "double exception to the rule of cell specialization." This makes them both doubly useful and exceptionally interesting.[14]

Stem cell technology offers not only to compensate for the losses inherent in cellular specialization—such as aging, disease, or organ failure—but to reverse them and introduce an economy of growth in perpetuity. Stem cell technology, therefore, is not only offering new, lucrative, and "exciting" ways to harness the productive powers of the cell: what is most interesting about stem cell technology is that it appears to offer a new means of *creating* them.[15]

Root and Stem

The language of stem cell science—with its emphasis on propagation, cultivation, and growing, seeding, and harvesting cell cultures—is explicitly agricultural, while the high-tech manufacture of such cultures requires the infrastructure of modern laboratories that are both capitalized (or often, like Roslin, semiprivatized) and industrialized (in particular through mass processing equipment such as PCR robots and gene sequencers). Like early agriculture, however, stem cell science is not being led by capital as an independent force, but, rather,

Human Fertilisation and Embryology Act 1990

CHAPTER 37

LONDON: HMSO

£5·85 net

Britain's stem cell bank is regulated under legislation introduced in 1990 to oversee the procurement, handling, distribution, and storage of human embryos which, inevitably, are acquiring value as a form of life-stock or biocapital. *Courtesy of Her Majesty's Stationery Office.*

as Marx said of the industrialization of agriculture, by "particular lines of production" that, like domestication, sometimes succeed but largely fail. Marx's example of sheep raising in sixteenth-century England is apposite, for it, too, relied less on particular lines of sheep than on a system through which sheep breeding was supported by droving routes, country fairs, and, above all, by the wool trade that made the raising of sheep into a reliably lucrative line of business.

Building on its distinctive historical prowess in livestock management, traditionally linked closely to experimentation in the life sciences, Britain has quickly become established as the world leader in stem cell technologies, or what might be called "life stock" industries. As a *Business Week* headline announced in April 2002, "In Stem Cell Research, It's Rule Britannia" (Capell 2002, 1). Both Germany and the United States, Britain's major competitors, are hampered by strong public opposition to the use of human embryos for stem cell research, although these are widely considered to be the most important resource for this area of scientific innovation. Canada, Spain, Australia, Israel, Finland, Denmark, and Sweden have plunged into the business enthusiastically, while China, Japan, South Korea, and Singapore also have burgeoning stem cell industries supported by a combination of public and private investment.

In their report on stem cell research, published in February 2002, the Select Committee of the House of Lords in Britain claimed that

> until recently it has generally been considered that in mammalian cells the process of differentiation is irreversible. However, it has been demonstrated in animals that it is possible to reprogramme ("dedifferentiate") the genetic material of a differentiated adult cell by CNR [cell nuclear replacement]. Following this seminal finding, many studies have also suggested that adult stem cells may have greater "plasticity" than previously suspected: they may be reprogrammed to give rise to cell types to which they normally do not give rise in the body. The potential of specialized cells to differentiate into cell types other than those to which they normally give rise in the body is little short of a revolutionary concept in cell biology. It has significantly increased the possibilities for developing effective stem-cell based therapies. (House of Lords 2002, 13)

In this description, *plasticity* and *reprogramming* are the main terms used to describe what is revolutionary about the CNR technique. Closely following scientific accounts of stem cell technology and nuclear transfer techniques provided by scientists from Roslin who testified before the committee, the report's description of the basic biological breakthrough behind stem cell research endorses the view that it offers radical new possibilities, emphasizing their therapeutic potential.

The House of Lords report, intended to guide national policy, offers a thorough consideration of stem cell research and concludes that it should be "strongly encouraged by funding bodies and the Government" in Britain (House of Lords 2002, 48). Research on human embryos is described as "necessary, particularly to understand the processes of cell differentiation and dedifferentiation," and the Dolly technique is strongly endorsed in the statement that "there is a powerful case for its use . . . as a research tool to enable other cell-based therapies to be developed" (48–49). The report recommends the establishment of a British stem cell bank to be "responsible for the custody of stem cell lines, ensuring their purity and provenance" (50) and concludes that existing mechanisms for regulation of research, and mechanisms for procuring informed consent from donors, are

A

BILL

INTITULED

An Act to make provision in connection with human embryos and any subsequent development of such embryos; to prohibit certain practices in connection with embryos and gametes; to establish a Human Fertilisation and Embryology Authority; to make provision about the persons who in certain circumstances are to be treated in law as the parents of a child; and to amend the Surrogacy Arrangements Act 1985. A.D. 1989.

The Human Fertilisation and Embryology Bill. As the social lives of human embryos become increasingly active, so too are their connections widened, so that today it is not unusual to hear detailed discussions about embryos, parts of embryos, and human embryonic cell lines on the evening news, or in the houses of Parliament. *Courtesy of Her Majesty's Stationery Office.*

sufficiently robust to accommodate the new developments posed by the area of stem cell research.[16]

Although the committee acknowledges that it was only able to give limited attention to the role of commercial interests in stem cell research, it did devote an entire section to this concern, acknowledging that its members "have been aware throughout that commercial interests could, and to some extent already do, play an important role in the development of such research" (32). They also acknowledge that "biotechnology is a growth industry," citing an Ernst and Young report that by the end of 2000 "the total value of Europe's publicly quoted biotechnology companies stood at 75 billion Euros, compared with 36 billion Euros a year earlier" (32). The committee adds that "according to a separate report, the United States, which has the largest number of companies in this field, market capitalisation of publicly quoted biotechnology companies fell over the same period (from $353.8 billion to $330.8 billion), but the number of public companies increased by 12.6%, and in the two years to June 2001 biotechnology stocks outperformed internet stocks on the Nasdaq index" (32). These references, along with acknowledgment that the United Kingdom "has by far the most public biotechnology companies" in Europe, and that "investor interest is considerable and evidently based on the assumption that future profits may be significant," confirm the extent to which economic growth in the biotechnology sector is viewed as a national priority. This is further underlined by references to China and Singapore, which "provide examples that deserve special mention":

> In China the government has encouraged a number of universities to invest heavily in stem cell research. In doing so universities have attracted not only public funds but investment by private companies like the Beijing Stemcell Medengineering Company. Leading Chinese researchers are often US-trained and have links with American laboratories. In Singapore, the Economic Development Board has provided initial finance for the Singapore genetics Programme; it is said that by 2005 some $7 billion dollars will have been invested in relevant research. In both China and Singapore there is concern with ethical issues but also an interest to maintain the competitive advantage gained by light regulation. (32)

Between the lines of this description emerges an acknowledgment of the intensely competitive economics of the global biotechnology sector, as well as a recognition of possible tension between "concern with ethical issues" and "the competitive advantage gained by light regulation"—an implication that foreshadowed ethical controversy about South Korean cloning programs in 2004 and 2005.[17]

In sum, Britain has successfully promoted its highly regulated but unusually permissive biotech research and development environment by emphasizing its stability, in large part due to high levels of public confidence in the government's ability to regulate developments in the life sciences, which in turn rely on a long history of public and legislative debate about reproductive biomedicine. This scene was largely set through the debates over IVF and embryo research that began after the birth of Louise Brown in 1978, and resulted in the establishment of the Human Fertilisation and Embryology Authority (HFEA) in 1990 (Franklin 1997b, 1999a; Strathern 1992b). Public confidence in the HFEA has been a key factor in Britain's ability to lay claim to a stable ethical and social, as well as legal, climate for investment, and the combination of high public trust and robust regulatory guidelines is regarded as a competitively advantageous recipe for long-term research and development, which the British government is keen to protect and maintain. That a technique developed by an agricultural research facility largely concerned with livestock breeding has in such a short time become one of the principal techniques of an emergent global biotechnology industry, which, because it increasingly relies on human, not ovine, embryos, is regulated under the aegis of a licensing authority established to oversee reproductive medicine in Britain, returns us to the theme of the complex hybridities, conjunctures, and mobilities that surround Dolly and the means of her creation. In turn, the complex social and legislative work of cultivating public opinion, and propagating venture capital investment, confirm the necessity of successfully passaging stem cells through more than one kind of culture medium.

Britain's First Lines

The opportunity to witness the propagation of cell lines for the first time in person made these agricultural analogies, and the coupling of agriculture and experimentation with life stock, even more vivid for me, and also evoked a new set of comparisons to the processes of settler colonialism, to which sheep also proved central. The following is an extract from my field notes after a visit to the laboratory in which Britain's first stem cell lines were made, in the spring of 2003. The week before I wrote these notes, the King's College cell lines were announced to the press with much fanfare as "the first in the UK." "Scientists Grow New Heart Cells in Test-Tube" proclaimed the *London Evening Standard* in the week before my visit, in mid-August 2003. In fact, scientists at King's College, led by the embryologist Sue Pickering and the neurobiologist Stephen Minger, had been highly successful in making human embryonic stem cells, using blastocysts donated from the Assisted Conception Unit at Guy's Hospital, for over a year. Pickering was granted one of the first two UK licenses to make human embryonic cell lines by the Human Fertilisation and Embryology Authority, which has overseen and regulated all embryo research in Britain since it was established in 1990, in February 2002.[18]

When I met Sue on the Tuesday following the announcement she was not happy about all the media publicity: "I hate all this press stuff," she complained. But she was very happy about the cell lines: there were more and more lines coming on very nicely in the lab.

> Tuesday 19th August, 2003, London
>
> Sue took me over to the Hodgkin building at about 3:30 to show me the lines. We walked past the porters at the main entrance and through two

sets of security doors before we got to the lab on the second floor. Nothing looks very "state of the art"—indeed some of the cracks in the old lime green terrazzo floor are as wide as a 5p coin.

I am introduced to Minal and Sara, Sue's research associates, in the lab as we go in. They are wearing white coats and I am fitted with one of Stephen's extras from the back of the door. The lab is very small, and could be an undergraduate teaching facility it is so basic. Minal has a square covered tray containing four sealed wells of stem cells under the microscope. She's new to the lab, and is quite pleased with the colony she has put through four passages successfully. It's from ESI, Alan Trounson's Singapore unit. A small vial of 6 colonies costs $6,000. It has been through 84 passages without differentiating and is being grown on an equally carefully characterized bed of mouse feeder cells ($500 per vial). This line is the control.

I have to put on white latex gloves and rinse them with disinfectant before I can approach the microscope. Sue does the same, expertly snapping her gloves up over the ends of her lab coat sleeves, which I imitate somewhat less successfully.

I have a look through the viewfinder and it is immediately obvious which are the stem cells, as they look like a flat bed of round amber pebbles surrounded by the stringier, more brachiated feeder cells. The surface of the stem cell colony is very uniform in color and texture, which is what makes it clearly "characterized" or differentiated. It has a unity distinct from its surrounding cells, and appears a bit like an island floating in a feeder sea. I can see at once why Minal is happy, and why this is a "successful" colony.

Sue shows me one of the lines they have made themselves, which looks quite similar, except that it has a kind of brown, furry mound in the middle. It is unattractive and looks like a mouldy pie. Sue explains this is what she calls "the dog's dinner phase," during which the colony forms a kind of brown scum on top. She calls it a "button," but it looks more like a patch of mildew, which, apparently, later disappears.

As I had experienced before in visits to the lab with Sue, it is striking what a difference it makes actually to see something for oneself (Franklin 2003d). My first thought was that the word *line* is very much at odds with what you see when you look at one. They do not look like "lines"; they look like ponds or landscapes. They look a little bit like the skin on the back of your hand when it is magnified, except there is no grain or apparent organization among the cells, which appear instead to be randomly strewn together, albeit with an overall coherence.

In my field notes I recorded my initial reactions to witnessing stem cells in person:

> Looking at a stem cell colony, you can see immediately what is changing about the "new biology." It is not about development and form in the older sense of the organism, but about parts. This is a biology that is about reassembly. It is about using the logic of the system or totality, but applying it to parts, which in turn are being used to make new "wholes." It is very important that the stem cell colony has a systematic integrity, that it is clearly differentiated, and well characterized. This stability and uniformity is "good form and function" for a cell line. But it must be singular and identical, and not developing new parts. It must be almost the opposite of the old developmental biology in that it should be both reproducing and staying the same—ironically so its "plasticity" can be harnessed. It should be continuously propagating sameness, in a process of perpetual nuclear fission. Separately controlling reproduction, differentiation, and development in this way is the most important form of biological control now, at the outset of stem cell propagation. Being able to direct the cells to differentiate "to order" will be the next step in this labour-intensive new field. That is how the cells in the colony will be put to work—by having their reproductive power redirected, re-instructed, and re-deployed. The total assemblage, of human cells, mouse feeder cells, plate, microscope, lab, scientists, university, etc. fuses cellular matter with all of the culture ingredients needed to sustain it as a productive, generative, system.

Later, writing up my notes, I thought more about the relationship between the idea of the colony and the idea of the line. Both of these, and the ubiquitous presence of agricultural analogies, made me think that stem cells are not only about the new biology but about how much it has in common with some of the oldest meanings of the biological, especially where it intersects the horticultural—which, interestingly, is where the term *clone* derives, from the Greek for "twig" or "stem."

The idiom of the line conveys the principle of unilinear, or one-way, descent—the defining vertical, and downward, orientation of the post-Darwinian biological, in which reproductive substance "codes" for new life through the process of genetic inheritance. The use of the term *colony* is biological, referring to a genealogical population or group. According the first definition of *colony* from the *American Heritage Dictionary*, it is "a group of emigrants or their descendants who settle in a distant territory but remain subject to or closely associated with the parent country."

The second definition emphasizes political control: "A region politically controlled by a distant country: a dependency." These two definitions together yield the biological definition of *colonial*, which means "living in, consisting of, or forming a colony," for which the example is "colonial organisms." Sue's colonies epitomize the definition from microbiology: "A visible growth of micro-organisms, usually in a solid or semi-solid nutrient medium." They are groups of cells derived from a common ancestor, laboriously brought under reproductive control and fed on a nutrient-rich diet to maintain successful propagation.

Significantly, the term *colony* derives both from the Latin *colonus*, for "settler," and from *colere*, for "cultivate." Colony in its biological sense thus refers both to the movement and spread of populations and their cultivation, or propagation. Propagation provides the horizontal axis of the stem cell economy, or culture, in that the basic idea is for them to multiply. *Propagation* is defined as "multiplication or increase, as by natural reproduction. The process of spreading to a larger area or greater number. Dissemination." Stem cells are ideally, then, both singular and multiple: the uniform singularity of their inheritance, or lineage, preserved as they multiply or spread.[19]

The propagation, or gardening, analogy became very evident when Sue showed me a large piece of the inner cell mass of a blastocyst that she had been using to seed a new line. The plump cell is literally planted into the feeder bed, which provides the archetypal Aristotelian "soil" to support its growth. Nothing appears to be happening just yet, but in theory this is the beginning of a new colony, or line. Sue is widely praised among the scientific community for her "green fingers"—a variant on the ultimate laboratory compliment of having "good hands"—referring to her proven talents at growing stem cells from scratch.[20]

In addition to her green fingers, Sue also has "magic hands": she is a master of the micromanipulator and can remove an intact inner cell mass (ICM) even from a collapsed blastocoelic cavity ("with per-

Wether sheep are the hallmark of a stratified sheep breeding system in which maximum use of the most fertile land to fatten terminal rams is achieved by moving sheep around, keeping flocks for specialised purposes, and maintaining a variety of sheep that can be cross-bred. This system extracts the most value out of both the soil and the animal, while also turning seasonal and climatic variation into an advantage. *Reprinted from Henry Stephens,* The Book of the Farm, *vol. 1, 2nd ed. (Edinburgh: William Blackwood and Sons, 1851).*

severance," she adds). She can both "see" and manipulate the component parts of early embryos using specialized pipettes the width of ultrafine wool. With these tiny "hands," she is able to dissect, remove, and transfer the entire ICM, full of primitive human embryonic stem cells, into a well coated with 0.1 percent gelatin and mouse embryonic fibroblasts (MEF). These are the so-called feeder cells over which the growing stem cell colony will be passaged.

The fledgling cell colonies, or epiblasts, are in need of almost constant care and attention, and most of them will fail to thrive. Initially, they are grown in culture medium for varying lengths of time, from two days to two weeks, and kept under careful observation. Colonies which appear to be producing stem cell–like morphology, usually near the center, are transferred to fresh wells, or "replated," and washed in a process that must be undertaken with extreme precision to a relentless schedule. When a colony has proliferated without differentiation into what appears to be a stem cell–like colony, it is carefully cut out of the surrounding endoderm with a glass needle and removed intact as a clump to be further propagated as a line in a new well, until it is large enough to be split again.

Candidate, or putative, lines are rigorously tested using numerous techniques to analyze their immunocytochemistry, proteins, and genes to confirm their purity and provenance. If they are consistent in their faithful reproduction to type, these lines will be considered to have been "immortalized" and will be candidates for being banked in a public storage facility. Eventually, once enough high-fidelity, immortalized lines have been banked, it will be possible for each line to be more fully characterized, through a process of comparison, that will enable each of them to acquire a specific identity linked to the properties making them suitable for specific uses, either clinically or for research.

Parallel Imprinting

Like Britain's famous system of stratified sheep breeding, the lines of the genealogical grid proper to the new era of biological control are neither bilateral nor closed. Hardy hill ewes such as the Scottish Blackface, who are "good at IVF," live in the rough northern country where they can survive better than any other breed, relying on their keen ability to transmit across generations an essential knowledge of their territory to protect themselves against the extremities of climate and terrain. These ewes are crossed with long-wool rams to produce half-bred ewes, or mules, who should be prolific, hardy, and quick to mature. In turn, the mules, who are moved down country into the richer pastures of the south, are crossed with a terminal sire to produce the so-called fat lambs for market.

At the heart of the stratified system lies the principle of transfer, whereby the maximum benefit can be eked out of the desolate hill country, or "rough grazings" that comprise one third of Britain's available pasturage and passed down through a system of crossbreeding that will amplify its value through a complex "stratified" breeding system that is based on a combination of diverse sheep, specialized flocks, and annual market cycles. The strength of the system is thus its ability to transform the diverse climates of the British Isles into a flexible breeding system that can be adjusted to maximize output by strategically controlling input through movement.

A rich mix of genealogical variety is essential to the flexibility and plasticity of Britain's stratified sheep breeding system, in which it is necessary to both maintain animals that are fit for purpose, while also introducing changes and adjustments to the system. In this system selective breeding involves adding and subtracting, but above all balancing, desired traits. *Courtesy of the British Wool Marketing Board.*

The modern British system of stratified sheep breeding builds on the older European transhumances, whereby sheep were walked on annual circuits that took advantage of seasonal changes in pasturage. According to Ryder (1983), sheep walks are very ancient, having been recorded in the writings of Pliny and Varro. The lines of transhumanant husbandry remain etched into the geography of contemporary Europe as railways and highways, along which sheep are still transported to higher mountain grazing in the summer months.

These ovine imprints on modern European soil are echoed in the crisscrossing cell lines of humans and sheep at leading agricultural facilities such as Roslin in a new genealogical grid of biological connection we could characterize as a primordial stock exchange. A different kind of stratified breeding system now links the banking of cell lines with the transgenic possibilities made available through techniques such as cell nuclear transfer. These offspring of the "era of biological control" comprise new forms of life stock that have been designed and harnessed to exploit the reproductive totipotency, and plasticity, of cellular functions previously unknown to science. The rich mix through which techniques of propagation and cultivation are combined with reproduction and replication maximizes the recombinant possibilities of both genealogical form and substance.

In this new genealogical *recombinatoria* the ties that bind are no longer strictly constitutional, inherited, or irrevocable, but are rather contingent, reversible, and plastic. An implosion of the practical and the speculative that is primarily agricultural in origin renders the new grids and banks of cell lines dense with commercial, reproductive, and industrial purpose. As new scientific frontiers, stem cells offer the possibility of colonizing a new genre of interior: the shared interior of the vital mechanisms linking humans, animals, and machines in new kinship configurations that reverse the order of substance and code—socially, genetically, and otherwise.

In her kinship with some of the oldest and newest definitions of capital, as in her novel relationships to sex, Dolly's identity as a clone can be seen to acquire added dimensions linking the stem technology used to make her to both ancient and contemporary ideas of stock. Against the immediacy of the many questions about how stem cells will be derived, manufactured, stored, commercialized, and applied, Dolly enables us to pose broader questions about, for example, the relationship of livestock to life stock in terms of how reproductive power can be controlled, managed, and commodified. For these reasons, Dolly constitutes a model organism for more than a specific technique, and a viable offspring of many genealogies. Through the lens of her dense historical embodiment of sheep breeding-crossed-with-laboratory science, she helps us recognize the inseparability of her commercial significance from both the complex legacies and uncertain futures of biocapital.

3 Nation

The cultivation of sheep and the manufacture of fleece have, from the earliest period of history, formed the most important branches of the agriculture and commerce of Great Britain. . . . The sheep and its wool were early and unequivocally acknowledged to be the foundation of national prosperity and wealth.
—William Youatt, *Sheep*

The sheep is, of course, an animal of many potentialities.
—Robert Trow-Smith, *A History of British Livestock Husbandry to 1700*

It may be that the average Britisher has a fondness for sheep, for we find them very numerous in our dominions.
—J. F. H. Thomas, *Sheep*

The historical relationship between sheep breeding, agriculture, industry, and commerce is as densely interwoven in Britain as anywhere else, but in few other countries has the role of experimental breeding—the ability to change the inherited constitution of sheep—been as prominent, or as promiscuous, in its links to other sectors of innovation, industrialization, and their foundational discourse of improvement. The viability and plasticity of British sheep has long been a source of their inestimable value to the British people, "the measure of national prosperity or calamity," in the words of William Youatt (1894, 2), the leading nineteenth-century sheep historian. In terms of nation—as in terms of sex, capital, and empire—sheep genealogies mix some of the oldest traditions of belonging to blood, soil, and country with their modern, contemporary, and future forms in ways that confirm both their ordinary and surprising connections to the identities, economies, and modes of reproduction they share with their human keepers.

It is estimated sheep first came to Europe from Asia and the Middle East over five thousand years ago, and in Britain, as in many other

The Wensleydale is a breed of long wool sheep developed by Richard Outhwaite in the Yorkshire Dales in the mid-nineteenth century. The breed is renowned for its stature, lean meat, scrapie resistance, and fine wool. Among the largest of the indigenous English breeds, it is a direct descendant of Bakewell's Dishley sheep and is now a listed rare breed. *Reprinted from William Youatt,* Sheep: Their Breeds, Management, and Disease *(London: Simpkin, Marshall, Hamilton, Kent, 1894).*

parts of Europe, their unique ability to survive under the most inhospitable conditions, in some of the most exposed and remote parts of the country, with little assistance from their keepers, has made them a highly prized source of wealth, as well as a definitive feature of the landscape. It is no coincidence that Dolly is a sheep, and it is equally significant that she is British, or more specifically Scottish, for she embodies the combination of medical, agricultural, and industrial values that gave rise to many other noteworthy Scottish inventions, including penicillin, the steam engine, the cotton reel, nude mice, interferon, and the syringe.[1] In sum, Dolly is as local, regional, and national as she is global an animal, for reasons that make her a very ordinary, as well as exceptional, sheep.

Many breeds of sheep remain distinct in Britain—more so than in any other country, and for geographical as well as economic, cultural, and historical reasons. The diversity of British sheep corresponds to the wide range of environments they inhabit across the British Isles and the many uses to which they are put: for wool, for meat, for milk, for breeding purposes, and to improve the land. Indeed, it is almost impossible to imagine the contemporary British countryside, its deforested green hills and picture-postcard dales, without imagining those hillsides covered with sheep. Sheep epitomize the countryside belonging to the industrial heritage that gave the British landscape its contemporary form, and they have remained of substantial economic importance for millennia, the foot and mouth crisis of 2001 only underscoring their ongoing centrality to the British economy.[2]

These histories of region, soil, and agricultural tradition are kept alive in the bodies of sheep whose breed names commemorate their origins in Dorset, Leicester, Lincolnshire, Hampshire, the Cotswolds, Suffolk, or Shropshire, each bloodline known for distinctive traits suited to its local environment, as well as those that can be reliably passed on through crossbreeding. Nowhere have these distinctive sheep traits and corresponding sheep-breeding lore enjoyed a healthier intercourse with the life sciences than in Britain, where Charles Darwin was famously influenced by the breeder's arts and the sheep farmer Joseph Banks served as president of the Royal Society.[3]

This chapter explores the changing national importance of sheep, emphasizing what is particular about sheep breeding in Britain, why it has been so important historically, and how sheep have played a role not only in the shaping of Britain itself but also of the British Empire. These observations are set against the long-standing importance of pastoralism to the world's earliest economies and the persistence of the importance of sheep as both an economic reserve and a sector of distinctive economic flexibility. In presenting this material, drawn from a wide range of nineteenth- and twentieth-century sources, two general points emerge as paramount. The first is the unique diversity of British sheep, which remains the defining characteristic of the British sheep population as a totality. This diversity is in part a consequence of British geography, but it cannot be attributed solely to that cause. It is also the ability to mix breeds to adapt to changing conditions and to preserve this mixture, something notable in both convention and rare-breed conservation in Britain even today.

Also notable, and the second point of emphasis, closely related to

Rare breeds tea towel. The conservation of rare breeds of sheep has become an increasing priority in Britain as a means of retaining as wide a range as possible of genetic diversity to draw upon for breeding purposes, but also as a living repository of British agricultural heritage. *Courtesy of the Rare Breeds Survival Trust.*

British sheep in New Zealand. Many of Britain's important economic, scientific, and cultural ties with its former colonies, particularly New Zealand and Australia, were established through sheep. British sheep remain the primary source of breeding stock for both countries. *Photograph by Sarah Franklin.*

the first, is the broad mixture of uses to which sheep have historically been put in Britain, distinguishing it from the majority of other leading sheep nations such as Spain, Australia, or New Zealand, where the economies based on sheep have been more single purpose, either for meat or for wool, and involving many fewer breeds. These changing uses of British sheep now include heritage preservation (rare-breed industries),[4] tourism (e.g., Herdwick sheep, the Lakeland Sheep and Wool Centre),[5] biotechnology (e.g., "pharming"), genomics (ovine DNA mapping), and medicine (sheep are the domestic animal most closely resembling humans in their reproductive and respiratory systems), as well as for the traditional uses of wool, meat, milk, cheese, fleeces, breeding, and land replenishment.

Sheep breeding in Britain is distinguished by a high degree of mobility as well as diversity. As noted in the previous chapter, the system known as stratified breed crossing generally involves a downward movement from the hardy mountain sheep of Scotland, Wales, and the north of England, which are crossed with larger, long-wool rams from the hill regions of central Britain, followed by another set of crosses with lowland sheep of the south. This system maximizes the value of otherwise uneconomical land in the remote northern regions by, in effect, passing down its value through successive crosses. The most common crosses, of purebred mountain ewes with long-wool rams (e.g., Swaledale crossed with Bluefaced Leicester), produce half-bred ewes known as Mules. These ewes produce terminal rams who are finished for the meat market and daughters who are reared and crossed with "down" rams from the south to produce prime-quality terminal lambs.[6]

This integrated system of crossbreeding depends on a complex network of specialized regional markets, such as the seasonal fat lamb markets that take place every spring. September is the season for agricultural and country fairs, which, since the Middle Ages, have played a critical role in dispersing sheep for crossing nationwide. As the agricultural historian Robert Trow-Smith writes: "The market and the fair were, and are, the veins of the livestock industry: without them the blood of pastoral production would cease to run" (1959, 226).

The idea of a market as the lifeblood of an industry takes on additional meanings in the context of livestock breeding as the goal of commercial profit through the control of reproduction makes of the bloodline a literal conduit of capital accumulation. Moreover, this bloodline is at once that of an individual animal (the commodity) and a much larger system (the means of reproduction). The wider system, of circulation (if we retain the sanguinary analogy), depends, in Britain more than in any other country, on the maintenance of a diversity of sheep types—both to maximize efficiency and to retain flexibility—in order to respond to market changes such as shifting preferences for meat or wool. Blood, soil, commerce, and capital thus shape animal bodies as embodiments not only of meat, wool, and other sheep products but of the mixed constitutions necessary to maintain, and adjust, the complex reproductive system that forms the basis of the stratified livestock industry.

As we have seen in the previous chapter, appreciating the reasons why Britain remains the home of the world's most diverse sheep population and is still—in the twenty-first century—the stud stock capital of the global sheep industry, frames the importance of more recent sheep-breeding innovations, namely Dolly, in terms of continuity with an industrialized agricultural past. This, in turn, helps to offset the hyperbole of radical novelty that, understandably but sometimes unhelpfully, so often accompanies discussions of cloning and biotechnology. Without understating what is indeed very radical and new about Dolly's conception, as outlined in chapter 1, it is entirely possible to situate what has come to be known as the Dolly technique within the wider frame of reference provided by her sheepish genealogy.

In this chapter, then, the questions of what makes a breed, what makes genetic capital, and, indeed, what makes a sheep take on additional dimensions that contextualize the Dolly experiment within the ancient and contemporary histories of sheep breeding in Britain. The ways in which the bodies of sheep have been reshaped, recapacitated, and selectively bred for manufacturing purposes is, after all, a very old story. So, too, is the effort to make of sheep distinctive forms of national economic prosperity, as evidenced not only by Australia and New Zealand but, earlier, by Spain, Germany, and France. Even before the birth of nations or national economies, sheep were critical to the economies of classical Greek and Roman society, to biblical definitions of wealth and human sovereignty over animals, and to Asian, Middle Eastern, and North African societies in the earliest periods of human presettlement (for a full review see Ryder 1983). In trying to foreground some of the broad themes that recur throughout the effort to harness sheep capacities for human ends—across several millennia and through initial domestication followed by various applications of the breeder's arts—we return to the twenty-first century somewhat more appreciative of the ways in which not only sheep breeding but also the emergent industry of stem cell manufacture recapitulate familiar cultural, historical, and economic themes.To be clear, the aim of emphasizing historical continuities by pointing to all of the ways people have always shape-shifted their sheep is not to diminish what is morally challenging, ethically uncertain, or troubling about Dolly. I do not mean to construct a defence of cloning in the humans-have-always-modified-nature vein of arguments used to support the genetic modification of plants, animals, and microorganisms by saying that it is nothing new, or, indeed, that it is neolithic. Rather, it is precisely because questions about the future of nuclear transfer technology matter as much as they do that their consideration should not be limited to a presentist frame of reference.

National Animals

The unique importance of sheep is well known for its historical role in the formation of ideas of Britishness, British industrial heritage, and the British countryside, but their ongoing contemporary significance is equally striking. For example, as part of a 1998 campaign to rebrand post-Thatcherite Britain, John Williamson, a design consultant from the firm Wolff Olins, proposed the theme of sheep. Selecting "three sheep representing complementary images of Britain—natural heritage and tradition; radical eccentric British culture and creativity; and [British] leadership in innovation, science and technology," Williamson was depicted in the national newspaper, the *Independent*, with a large poster board containing three different iconic images of sheep. In the central image, a flock of sheep are crowded into a country lane, accompanied by a shepherd, in a classic scene of British country life. The image evokes a sense of age-old rural agricultural traditions in the same generic vocabulary as do hedgerows, dry stonewalled pastures, or thatched farmhouses that today belong equally to the tourist and the agricultural economies. Although the image is pastoral, and sheep are more commonly figures of fun than of national pride, no account of British sheep breeding, as we shall see below, could ignore the substantial innovation necessary to establish and maintain its uniquely complex methods of livestock production and management.

In the second image, the British conceptual artist Damien Hirst's Turner Prize–winning installation of a lamb preserved in a glass box filled with formaldehyde, entitled *Away from the Flock*, gestures toward a different tradition of innovation and creativity in the form of eccentric British radicalism. The rise of the so-called YBAs, or Young British Artists, in the 1980s constituted one of the key elements, along with the worldwide success of pop music bands such as the Spice Girls and Oasis, in the short-lived celebration of New Labour's "Cool Britannia" in the 1990s. Hirst's controversial piece extends his fascination with preserved animal specimens, but it carries a more unusual religious connotation as it contains a sacrificial lamb of sorts. Unlike the flock of sheep shown beside it in Williamson's sheepscape, Hirst's lamb is alone, a captured individual away from the flock. That it is dead emphasizes the tragedy of its isolation, while its typically frolicsome lamblike pose adds a sense of lost innocence

DESIGN

DESIGNS ON BRITAIN'S FUTURE

Eighty of our top creatives were asked to pick an object that represents what's good about Britain. Jeremy Myerson (above) was there

PAUL SMITH
FASHION DESIGNER
'This demonstrates how to mix tradition with our more recent achievements and successes'

SIMON ALLFORD
ARCHITECT
'The Routemaster is an image of civic design, mobility, mass production and therefore design – from the curve of the bonnet to the number of the bus route'

JEREMY MYERSON
JOURNALIST
'This statuette of John Barnes in an England football shirt is a symbol of Britain because it represents multi-cultural Britain and our sense of fair play'

MARY LEWIS
GRAPHIC DESIGNER
'My image of Britain is a fertile place, alive with ideas – something that must be nurtured'

JOHN WILLIAMSON
DESIGN CONSULTANT
'Three sheep representing complementary images of Britain – natural heritage and tradition; radical eccentric British culture and creativity; and our leadership in innovation, science and technology'

NIGEL COATES
ARCHITECT
'This Vivienne Westwood shoe sums up our ability to invest an object with cultural resonance, in this case the irony of the nurse as an image of courage, power and sex as well as compassion and service'

JOHN WILLIAMSON
DESIGN CONSULTANT

'Three sheep representing complementary images of Britain – natural heritage and tradition; radical eccentric British culture and creativity; and our leadership in innovation, science and technology'

Williamson's sheep-scapes. Sheep have long been important culturally as well as economically to many countries, and in this rebranding exercise are chosen to represent Britain as creative, ingenious, artistic, and innovative as well as the product of distinctive rural traditions.

or of stilled life—its silent, immobile form reproducing the terminal reality of the lamb destined for the slaughterhouse. Lifelike and intact, the lamb is at once embalmed in a tomblike enclosure and yet remains oddly animate in its clear amniotic environment, as if captured in a photograph mid-frolic. This liminal state between life and death—open to the suggestion of reanimation, and thus reincarnation—underscores the strong associations linking sheep to Christianity, while simultaneously invoking the medical and zoological associations with gross anatomy, taxidermy, and "the baby in the bottle."[7]

Innovation in a different sphere of science is signified by Williamson's third image, of the bookended duplicate image of Dolly from the cover of the *Economist* in March 1997 announcing her birth. Sheep such as Dolly directly extend the breeding innovations pioneered by British agriculturalists such as Robert Bakewell, who created the first modern sheep through systematic in-and-in breeding in the mid-eighteenth century at his Leicestershire farm. Dolly, too, belongs to a heritage of innovation in animal breeding, and in the life sciences in general, for which Britain is historically celebrated and uniquely accomplished. From Charles Darwin and Alfred Wallace, or Francis Galton and Thomas Huxley, through James Watson and Francis Crick, Robert Edwards and Patrick Steptoe, and most recently Keith Campbell and Ian Wilmut, a legacy of radical experimentation and intensive scientific investigation of reproductive biology distinguishes Britain as a nation highly tolerant of the manipulation of life in the name of scientific progress and economic growth (see also Graham 2000).

In these three images of sheep, a common domestic farm animal, Britain is represented as innovative, pioneering, radical, trendy, quaint, traditional, and eccentric. Sheep are depicted in Williamson's selection as a singleton, as duplicates, and as a flock. The images make reference to the realms of art, science, and agriculture, and in them sheep can be "read" as ordinary, exceptional, sacred, economical, and tragic. All of the sectors referenced by sheep in this rebranding exercise are meant to suggest desirable qualities of British character (radicalism, individualism, eccentricity, inventiveness) and to point toward their economic importance to Britain as a competitive world power through its leadership in science, agriculture, and the culture industries.

Robert Bakewell's selectively bred New or Dishley Leicesters are often described as the first viable offspring of modern industrial livestock breeding, and epitomise the combination of enterprise, calculation, agriculture, and the breeder's arts that remains central to its organization today. *Reprinted from William Youatt,* Sheep: Their Breeds, Management, and Disease *(London: Simpkin, Marshall, Hamilton, Kent, 1894).*

As an ensemble, these sheep offer images of nature, culture, and industry woven together in sheep's wool to suggest the fabric of a nation in which even the animals are man-made. In each case, the sheep are instruments of redesign or reconception—or literally both in Dolly's case. The country sheep all look alike because they have been bred to type, and although they epitomize the countryside, it is the industrialized, deforested, hedgerowed countryside—which constitutes a built, and carefully engineered, environment—they occupy. The layering of road systems onto sheep drovers' trails maps the co-influence of domestication on people and animals, binding them securely together into age-old pastoral formations.[8] Hirst's lamb, as a piece of conceptual art, powerfully reinvokes these associations in dramatically simple form, in which the encasement of the animal recapitulates its history of enclosure and domestication, as well as the intimacy of that relation to its "masters." Dolly, in a reverse effect of nontransparency, must have her image duplicated to become visible as a clone.

British Breeds

Thus, as both Williamson and Hirst demonstrate, the ability of sheep to represent the nation of "Beefeaters" is not surprising given that animal's primary, if often overlooked, economic importance for centuries—not only to Britain's regional, domestic, and national economies but to her expansion as a global empire from the seventeenth century onward, the industrial period with which colonialism was inextricably intertwined, and the modern era in which Britain remains one of the world's leading sheep markets.[9] As late as the 1950s, the British agricultural historian J. F. H. Thomas was still able to claim in his plainly entitled *Sheep* that "a surprising fact is that about one third of the world's sheep are to be found in the British Empire . . . and we produce one-half the world's wool from one-third of the world's sheep since we have in the British Empire a preponderance of wool producing types" (1955, 11). While Australia and New Zealand may be the world's leading sheep producers in terms of the percentage of their economies devoted to sheep farming, the British system remains the primary source of breeding stock still today, due to the exceptional diversity of sheep breeds and the elaborate system of crossbreeding used to maintain their distinctiveness.[10] As Thomas notes in one of the epigraphs to this chapter, "It may be that the average Britisher has a fondness for sheep, for we find them very numerous in our dominions" (1955, 20).

Much speculation surrounds the movements that first brought sheep to Britain, and, for that matter, the question of how the ancestors of the first domesticated sheep evolved, presumably in the Neolithic period and, it is speculated, probably in southwest Asia or what is now Iran.[11] Sheep, according to the *Encyclopaedia Britannica*, "belong to the family of hollow-horned ruminants or *Bovidae* (q.v.)" that "pass almost imperceptibly into goats." They were initially classified by Linnaeus in 1758. Four branches of the sheep family can be distinguished by their number of chromosomes—fifty-two, fifty-four, fifty-six, and fifty-eight (goats have sixty)—and are commonly divided into four groupings: the Urial (*O. vignei*) in Iran and India, the Argali (*O. ammon*) in Central Asia, the Mouflon (*O. musimon*)

Numbers of persons employed in the woollen and worsted industries by county, 1901. *Reproduced from J. H. Clapham,* The Woollen and Worsted Industries *(London: Methuen, 1907).*

in the central islands of the Mediterranean, and the Bighorn in the Rocky Mountains of North America (*O. canadensis*). William Youatt, the most highly acclaimed nineteenth-century authority on sheep, described "the zoological character of sheep," following George Cuvier's categorization of the order Ruminantia, as "having teeth in the lower jaw only, opposed to a callous substance in the upper jaw; six molar teeth on either side, and the joint of the lower jaw adapted for a grinding motion; four stomachs, and these, with the oesophagus, so constructed that the food is returned for the purpose of rumination; long intestines not cellated" (cited in Youatt 1894, 1). Within the order Ruminantia the "Tribe *Capridae*" was distinguished by permanent horns formed by "annual ringlets at the base" and dividing the surface of the horn "into a succession of irregularities or knots" (Youatt 1894, 1). These animals, including goats, were characterized by being "adapted for springing or swiftness" and lacking "any canine teeth in the mouth" and were distinguished from the genus *Ovis*, from which the familiar domestic sheep is descended. The genus *Ovis* contains, according to Youatt, "the *Ovis Ammon*, or Argali; the *Ovis Musmon*, or Musmon; and the *Ovis Aries*, or Domestic Sheep" (1894, 1).

The American author and sheep breeder J. F. Walker, writing in 1942, claimed that

> early breeds of sheep fell largely into three classes—and oddly enough could be classified by the lengths of their tails. There were long tailed, medium tailed and short tailed sheep. The long and short tailed breeds were primarily meat and milk producers and their wool was either a mixture of short, fine undercoat and long, hairlike outer coat, or even almost without wool or hair. It is from the medium tailed group that most of our modern breeds have descended. (Walker 1942, 9)

The Texan sheep historian John Ashton, writing in 1943, agreed that it was very difficult to determine the precise ancestry of modern domestic sheep, repeating the oft-encountered claim that "nobody knows with any degree of exactitude when the sheep first came on the scene" (Ashton 1943, 9). Drawing on the earliest archaeological evidence (the remains of domestic sheep were first discovered by archaeologists on a Swiss lake in 1861) and the speculations of other sheep historians, Ashton suggested a mixed ancestry that proved consistent with many other accounts. He claimed that "it is quite likely [sheep] were brought into Europe by two distinct and different routes. The Swiss lake-dwellers of early Neolithic times . . . possessed the 'peat-sheep,' which may have come from Asia via the Danube. This . . . may explain why some authorities state that the Asiatic Urial (*O. vignei*) or the Steppe sheep (*O. arkal*) are part ancestors of our modern breeds" (Ashton 1943, 9).

According to many sources, the ancestors of domestic sheep came from several distinct lines of descent. Writing in 1913, the famed agricultural zoologist R. Lydekker speculated that *Ovis aries* is itself a mixture, a "complex type" derived from more than one line of ancestry.

> The ancient sheep [referred to in the Bible] were almost certainly of Eastern origin, and thus derived, in all probability, from one or more Asiatic wild species; but it is also quite probably that the Prehistoric inhabitants of Europe tamed the wild mouflon, which although now restricted to the islands of Sardinia and Corsica, in former times probably enjoyed a wider distribution. If this be so, European domesticated sheep represent a complex type, derived from at least two totally distinct wild sources. (Lydekker 1913, 1–2)

Walker similarly suggests that "many consider the Mouflon or wild sheep of Europe to have played a very important part" in the evolution of early improved breeds of sheep (Walker 1942, 11). The Mouflon, which, as mentioned by Lydekker, is largely found in Sardinia and Corsica, so closely resembles the Argali of Asia and the Rocky Mountain sheep of North America that "by some, all three of these are considered identical" (Walker 1942, 11).

First Sheep

In contrast to the uncertainty surrounding the precise evolution of sheep in Europe, or the classification of sheep types ancestral to the modern domesticated breeds, there is ubiquitous agreement as to the importance of sheep to human societies from time immemorial. As Walker summarizes: "Sheep have been closely associated with civilization since its first history. The first record of livestock in the Bible was a reference to sheep. Ancient Egyptians perpetuated sheep through drawings and carvings in their oldest monuments, and even mummified bodies [of sheep] have been found. The early lake dwellers of Switzerland, the first inhabitants of Greece and Rome and the tribes of the Barbary States of North Africa, all had sheep as far back as records extend" (1942, 9).

It is today speculated that *Ovis orientalis*, the Mouflon or Asiatic Mouflon, "was most probably the ancestor of all domestic sheep as well as of the European Mouflon (formerly called *Ovis musimon*)" (Clutton-Brock 1999, 70, also see Kinsman 2001, 9). DNA analysis has confirmed a second ancestor for *O. aries*, although it has never been found, and is unlikely to have been a feral British sheep. The high percentage of hornless ewes among existent populations of both the European Mouflon and the ancient feral breeds of sheep in Britain, such as the Soay or Saint Kilda sheep from the Outer Hebrides, is seen as a remnant of early domestication, as it constitutes a trait rare among undomesticated breeds.

As Juliet Clutton-Brock has noted in her influential account of the domestication of mammals, there exist surprisingly few animals that have been successfully domesticated "despite more than ten thousand years of association with humans as the dominant species" (1999, 9). According to Clutton-Brock, "a domestic animal can be defined as *one that has been bred in captivity for purposes of economic profit to a human community that maintains total control over its breeding, organization of territory, and food supply*" (1999, 32; original emphasis). Most animals will not tolerate total control of this kind. According to Francis Galton, whose definition of domestication was once considered definitive, six necessary conditions exist for the potential domesticate: "1, they should be hardy; 2, they should have an inborn liking for man; 3, they should be comfort-loving; 4, they should be found useful . . . ; 5, they should breed freely; 6, they should be easy to tend" (qtd. in Clutton-Brock 1999, 2). Sheep amply fulfil all of these criteria, and they are assumed to be the first domesticated livestock animal, followed later by cows, pigs, and horses.[12] Unlike deer, for example, which have never been fully domesticated, sheep will tolerate containment. They can be "bunched up together in compact groups, often of both sexes, and they will even flourish better when crowded together" (Clutton-Brock 1999, 73). They have a well-defined social hierarchy, led by a single dominant individual, which makes them easy to tend and establishes a strong basis for animal-human communication.

The results of domestication in sheep have produced numerous physical changes over time, most notably a change in body size (initially domesticated animals tend to be smaller, but later the influence of selective breeding may have the reverse effect), a change in coat composition and color (with a tendency to favor lighter, woollier fleece), and a trend toward hornlessness (polled, or hornless, sheep were clearly selected for, particularly in the female animal).

In sum, the domestication of sheep may be considered an extension of human sociality to include specifically suited animals, of which there are few. In noting that "domestication begins with ownership" Clutton-Brock emphasizes that "incorporation into the social structure of a human community" requires that animals "become objects of ownership, inheritance, purchase, and exchange" (1999, 31), as well as being under the total control of humans. Domestication thus unites social control of animals with biological alterations to their constitution, extending "ownership" beyond *possession* to *creation*. The creation of animal populations that live in reproductive isolation from their wild ancestors, are physically contained, and whose breeding is controlled and directed, can thus be seen as one of the oldest human technologies, and one in which social and biological forces are inextricably intertwined.[13]

However, as noted earlier, and as is emphasized in the work of Haraway on companion species (2003), or in Helen Leach's perceptive insights into how the domestication process shapes humans in ways that may not be clearly anticipated (2003), sheep are both animals that epitomise domestication as subordination, while also being animals who have perhaps had a more dominant effect in shaping human societies than any other animal. In the same way that roads and turnpikes were eventually built along the ancient lines of sheep walks, and the dietary needs of sheep shaped eighteenth-century agricultural machinery, so the two-way features of the domestication process can clearly be seen in this useful livestock animal, so often associated with passive acceptance of their condition. As the anxiety and attention surrounding Dolly demonstrates, she matters because humans can be cloned, not only because sheep can, and for this reason alone she is already "one of us."

Antediluvian Sheep

Reference is often made to the biblical importance of sheep, both in terms of their early import on human society and their symbolic significance as embodiments of purity, innocence, and righteousness. As Lydekker comments, "The Biblical record indicates the importance of sheep to the ancient Hebrews, and also teaches that these animals

Today used largely as an ornamental sheep, and for their multi-colored wool, Jacob sheep are a small, primitive breed, often sporting four horns. They remain closely associated with their biblical role as the leading Old Testament sheep. *Courtesy of the British Wool Marketing Board.*

—especially as represented by lambs—were regarded as the emblem of purity, innocence, and righteousness, while their near relative, the goat, occupied precisely the opposite position" (1913, 1–2). Writing of "the antediluvian sheep," Youatt notes that Abel became a keeper of sheep, while Cain became a tiller of the ground, and that as their sacrificial acts of worship, Cain "brought of the first fruit of the ground an offering unto the Lord," while Abel "also brought of the firstlings of his flock, and of the fat thereof" (1894, 7). Abraham and Isaac were "the proprietors of numerous flocks in the East, and which the Arab and Tartarian shepherds continue to the present day" (Youatt 1894, 9). In addition to clothing, meat, milk, and fat, sheep, according to Youatt, "furnished the antediluvians not merely with coverings for their bodies, but also for their movable habitations," that is, the tents of early nomadic pastoralists, "the ancient shepherds," descendants of Jabal, "the father of such as dwell in tents and have cattle" (1894, 9). According to the Bible, "the flocks and herds of Abraham and Lot were so great that the land was not able to bear them," and "Job had 14,000 sheep as well as oxen and camels" (qtd. in Youatt 1894, 9).

In sum, sheep—often referred to as cattle—were "the chief possession—the almost only riches of the people" (Youatt 1894, 9). According to the Bible, sheep were used as tributary payments to patriarchs and kings, as sacrificial offerings for temple dedications and ritual expressions of piety, and for feasting and popular rejoicing. Sheep were a symbol of power and status, taken as bounty by victorious armies in battle, and used for barter to purchase land as well as bride wealth in the arrangement of marriages. Describing a transaction in which Abraham purchases a well with sheep, Youatt emphasizes that these animals were, in effect, a form of money: "Abraham effected a regular and freehold purchase of the well and the ground in which it was dug at the price of these seven ewe lambs. They were the money which he paid for this spot of ground; or 'they were that commodity of known value and general demand which stood in the stead of money,' proving how universally [sheep] had spread, and how generally their value was acknowledged" (Youatt 1894, 15). Elaborating his claim that sheep were, in effect, the first commercial funds—or even capital—with reference to classical as well as biblical sources, Youatt adds in a footnote that

> wealth used to be estimated by the number and quality of the cattle. *They were the principal instruments of commerce.* Thus we read in Homer of a cauldron being worth twenty sheep, and a goblet worth twelve lambs, &c. These animals were *the means by which exchange or commerce was originally carried on.* The proof of this is convincing enough, as well as very singular. The word which signifies the exchange of one kind of goods for another is derived from the Greek for a lamb. A wealthy person is called "a man of many lambs," and two rival brothers are represented by Hesiod as fighting about "sheep," that is, the property of their father. (Hunter, qtd. in Youatt 1894, 15–16; emphasis added)

The Bible is also the source of the first accounts of the improvement of sheep, in the well-known tale of Jacob and his flocks. Betrothed to Rachel (sometimes referred to as the first shepherdess), the daughter of the wealthy but manipulative patriarch Laban, Jacob labors as a shepherd for seven years, receiving no wages, but offering dedicated service. In a deal struck to achieve a kind of bride-wealth purchase of Laban's daughter, Jacob agrees to remain for another seven years in exchange for the privilege of keeping any sheep and goats that are speckled, or ring-streaked, as are contemporary Jacob sheep.

These sheep, "exceptions to the general colour, a brown or dingy black—the sportings rather than the regular production of nature—should be considered as the wages of Jacob" (Youatt 1894, 16). According to Genesis, Jacob brought about an increase in these speckled or ring-streaked "sportings" through the influence of setting before his ewes a row of "pilled," or peeled, rods, which, exercising a visual power over the female sheep at the moment of conception, yielded similarly streaked or "particolored" lambs:[14] "Jacob therefore took him rods of green poplar, and of the hazel and chestnut tree, and pilled white streaks in them, and made the white appear which was in the rods, and he set the rods which he had pilled before the flocks in the gutters in the watering troughs when all the flocks, male and female, came to drink" (Genesis 30:37, 38). By "setting the faces of the flocks toward the ring-straked" at the time of their mating, Jacob increased the number of ring-straked lambs that dropped, which, in turn, he used to produce more of their kind. As a result, Jacob eventually took control not only of Laban's flock but his wider assets, at the same time contributing to a consolidation of their whiteness. As Youatt speculates freely, "a selection of those that had the most white about them" yielded, at length, a fleece that was "purely white" (1894, 18), noting that by the time Solomon composed his famous song, the "pure white fleece" had become sufficiently proverbial that he could describe the whiteness of his mistress's teeth by comparison to a flock of sheep just come up from the washing (Song of Solomon 4:2), as David also compares fleece to the whiteness of snow (Psalm 147:16).

Roman Sheep

The influential Victorian polymath Youatt disagreed with the view that sheep arrived in Britain from a mixture of Asian and European routes, patriotically favoring the hypothesis that modern domestic sheep in Britain were descended from indigenous breeds inhabiting the British Isles long before the Roman conquest. In contrast, the postwar agricultural scholar Robert Trow-Smith, in his exhaustive two-volume history of British livestock husbandry, followed the archaeologist V. Gordon Childe in asserting that "it is generally agreed that no wild prototypes of the sheep survived in Britain into Neolithic times, [and that] their immediate origin cannot be in question: the first pastoralists brought them with them" (Trow-Smith 1957, 7).

He states that "it is certain that [the sheep] was introduced in its domesticated form, for no possible wild prototype existed here" (1957, 17), but adds that the evidence available is very thin: "Neither the excavator offering his own inexpert opinion nor the zoologist who may be consulted upon the animal remains from a site has found himself able to draw more than somewhat hazy deductions from the evidence before him. There is, therefore, no very exact picture to be painted of the Neolithic and Bronze Age type—or types—of sheep" (1957, 17). Neolithic pastoralists, assumed to have made their initial settlement in the British Isles on the Salisbury plain around 2500 BC, would have brought with them "a slight, small deer-like animal similar to the Soay sheep of today" (Trow-Smith 1957, 7).

Between the gradual increase in the population of sheep in the Neolithic period and the early Iron Age expansion of initial Roman settlement, which saw sheep numbers reach parity with those of cattle, significant ecological changes took place in the British Isles, including extensive deforestation, especially of the lowland downs, enabling sheep to occupy more land. Sheep were not only capable of producing meat, milk, and wool; they could also increase soil fertility, all the while requiring less water, less protection, and less food than cattle. "All in all, the sheep was a more productive, a more manageable and a more easily satisfied animal than cattle," Trow-Smith summarizes (1957, 36).

The first great expansion of British sheep came in the latter half of the Roman occupation, during the third century AD, largely due to the increasing importance of British wool as part of the imperial Roman textile industry. Sheep husbandry was of significant importance to the Roman Empire and constituted a highly advanced practice within Roman society, as it had been to the Greeks, from whom many of the more elaborate techniques of rearing, tending, and breeding were inherited. As Trow-Smith notes:

> The Roman nearer home practiced an advanced sheep husbandry. The ordinary breeds—those of Apulia, Calabria and Cisalpine Gaul—were improved by careful selection and were managed with much skill; and the highly prized Tarantine breed, concentrated in Southern Spain, were stalled and hand-fed (in winter on barley and split beans) and had their fleeces protected by a hide jacket—hence their name of *ovis tectae*. The Greeks in their time had been equally attentive to their flocks, rugging

> the sheep in winter, growing crops of clover, trefoil, Lucerne and vetches for their keep, yarding the stock at night, shearing them twice a year, and bathing them in salt as a preventative against mange. (1957, 38)

According to Trow-Smith, sheep breeding began to expand in Britain as a direct result of Roman occupation: "The appearance of weaving and wool in Britain appears to have coincided with the rise of the prehistoric British sheep industry to importance in the late Bronze Age" (1957, 19). The British sheep historian and wool industrialist Kenneth Ponting agrees that "white wool had become ever more in demand, particularly in Roman times, and the change from colour to white may be regarded as the most obvious result of domestication and proof, if any was needed, that from late Neolithic times onward the wool producing properties of sheep have been paramount. (Ponting 1980, 11). Ponting's view may in part reflect his influential position in the British wool industry, as other commentators frequently emphasize the ongoing importance in Britain of the sheep as a multipurpose animal, providing milk for cheese manufacture and fertility for marshland soils through their droppings and treadings. Detailed accounts from the so-called Domesday records show, for example, that according to the census ordered by William the Conqueror in 1066 there were as many as 92,000 sheep in Norfolk alone, providing a range of services and products for their keepers that were likely to be at least equal in value to their wool-producing capacity. As Trow-Smith notes, many of the English place names dating from the Middle Ages end in "-scap" or "-sceap," denoting their connections to sheep: "English place names, many of which probably provide earlier evidence of land use than even the very earliest of literary records, attest not only its presence but also its pre-eminence in local husbandry in a score or more of settlements whose names are compounded with sceap alone" (1957, 59). Indeed, Trow-Smith claims that "there is some reason to suppose that, even in the twelfth century before the great rural industry of wool production got into swing, the cow took a secondary place [to the sheep] in the pastoral scheme of things" (1957, 93). Thus, although the wool-producing qualities of British sheep may have been their most valuable market asset through the Middle Ages, as it had been also in the period of Roman occupation, the value of sheep to the developing British economy was far more diverse, reflecting its importance as a multipurpose animal integral to the development of land and soil, milk and cheese production, and the source of meat as well.

In sum, Britain was colonized by the Roman army and its sheep, and the early economy of Britannia was centred on wool. This pattern has persisted to such an extent that the chancellor of the Exchequer continues today to sit on the so-called woolsack in the British Parliament, at one time presiding over a quarter of the globe, as well as over an economy to which sheep proved equally essential at home and in the colonies.

The Wool Boom

Sheep and wool not only formed the basis of Britain's earliest economic and trade relations but also initiated the most significant changes in their form. The late medieval period of sheep husbandry in Britain is claimed by many historians to mark the first major rationalization and industrialization of wool production, through what Trow-Smith describes as "an unprecedentedly logical system of land use" (1957, 131). As noted both by Karl Marx (1972) and Fernand Braudel (1979), sheep farming in Britain provided one of the earliest and most important chapters of large-scale land management along the lines of modern industry and agriculture. The ramifications of this period are, indeed, inextricable from the reasons Britain later became the seat of the industrial revolution, and the current prioritization of biotechnology in Scotland owes much to its long history of innovation and rationalization of animal husbandry, particularly sheep breeding. As Trow-Smith claims at the outset of a chapter devoted to "the great industry of mediaeval sheep-keeping," its influence

> extended into the very heart of national policy and finance, into the contemporary and future influence of innumerable ecclesiastical bodies and lay families, into the international balance of power, and into the livelihood of the towns and countrysides of remote nations. The quality, price and quantity of British wool could bring prosperity or ruin to the weavers of Flanders, the flock-masters of Spain, the merchants of Italy; heretofore the well-being of a farm animal on an English common concerned only the peasant who owned it and his family. The change from an individual, self-subsistent stock husbandry, such as had been practiced in Britain for nearly 4000 years, to a vast network of industry and

commerce founded upon a humble sheep on this same common was as immense as any in history. (1957, 131–32)

This "immense" historical change, in Trow-Smith's view, was, like other important transformations in the history of British sheep breeding, one of kind as well as degree. The change in kind, referred to by Marx and Engels throughout their accounts of the emergence of British industrial society, was from a largely regional organization of sheep-related economies to national and international networks of interconnected market systems. On the back of the humble sheep, as the Australians were later famously to claim of their economic development, the British commons became linked to "remote nations" in a way that would directly alter "the balance of international power," confirming these farm animals' significance as the "heart of national policy and finance."

According to E. Carus Wilson (1952, 355–429), Britain's early role as a wool-manufacturing nation was as a supplier of woolen cloth for export, with the wool largely being woven in large urban centers and shipped abroad, mainly to Italian, French, and Flemish buyers who made it into clothing. As British wool began to be increasingly highly taxed overseas, and as new methods of woolen cloth production became more industrialized through innovations such as the fulling mill,[15] woolen textile manufacture moved out of the large cities and onto small farms and mill estates with access to clear, soft running water and advantageous weather conditions for spinning, such as the damp and misty climates of the Pennines, Lancashire, and West Yorkshire. Trade also began to soar in the West Country, Norfolk, and East Anglia, as well as in Wales, and the textiles of Herefordshire became so valued as to be described as "Lemster Ore" (Ponting 1980, 16). Initially, the wool boom was dominated by the influential Cistercian monasteries, with their large manorial flocks and well-established trade links. The later emergence of an economy based on privately owned flocks, the result of the breakup of some of the largest estates in the late fourteenth and early fifteenth centuries, led to increased competitiveness in the production of the largest possible wool clips for export.

Sheep breeding during this period would clearly have been oriented toward maximizing wool production, and the emerging prominence of the Merino sheep outside of Britain, which would eventually dominate wool production with its far superior fleece, was also directly

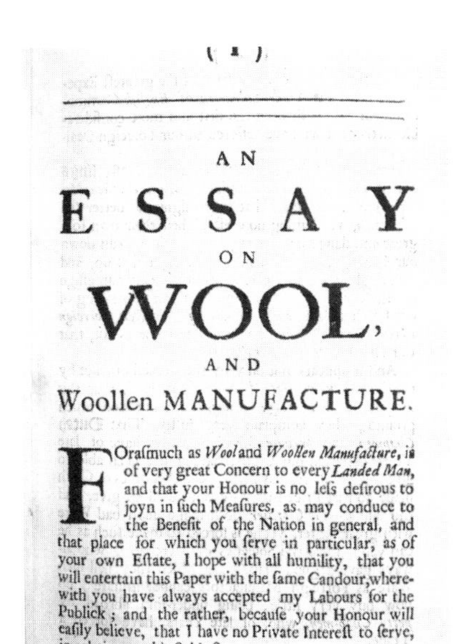
(1)

AN

ESSAY

ON

WOOL,

AND

Woollen MANUFACTURE.

FOrasmuch as *Wool* and *Woollen Manufacture*, is of very great Concern to every *Landed Man*, and that your Honour is no less desirous to joyn in such Measures, as may conduce to the Benefit of the Nation in general, and that place for which you serve in particular, as of your own Estate, I hope with all humility, that you will entertain this Paper with the same Candour, wherewith you have always accepted my Labours for the Publick; and the rather, because your Honour will easily believe, that I have no Private Interest to serve, in relation to this Subject.

B It

By the seventeenth century wool and woollen manufacture was sufficiently central to "the benefit of the Nation in General" to be described by the mercantile reformer Josiah Child as "of very great Concern to every Landed Man." Volume printed for Henry Bronwicke in London 1693. *Reproduced from the Rare book Collection of the London School of Economics and Political Science.*

related to the burgeoning global demand for wool. However, the value of sheep in Britain, unlike that of any of the countries dominated by the Merino—today the world's most numerous sheep—has always fluctuated between its wool-producing capacity, which at times becomes paramount, and other of its capacities, or more accurately, the value of its *mixed capacities* as an animal.

Two major changes in the seventeenth century would return British sheep breeding to a less wool-driven economy: one was the end of Britain's wool boom resulting from a combination of a government ban on the export of British wool (to satisfy the demands of British clothiers); the other was the increasingly widespread acknowledgement of the superiority of Merino wool over even the finest British long-wool products. The problem of how to increase the domestic supply of Merino wool could have been solved by importing flocks of Merino sheep from Spain, as was eventually advocated by the explorer and naturalist Joseph Banks, who first brought Spanish Merinos to George III's farm, at what is now Kew Gardens in London, in 1802 (Carter 1964). However, the Merino never became well established in Britain, and instead became the basis for Britain's colonial sheep and wool trade with Australia—with consequences that are explored in the following chapter.[16]

Merino sheep, often said to be derived from fine-wooled Tarentian

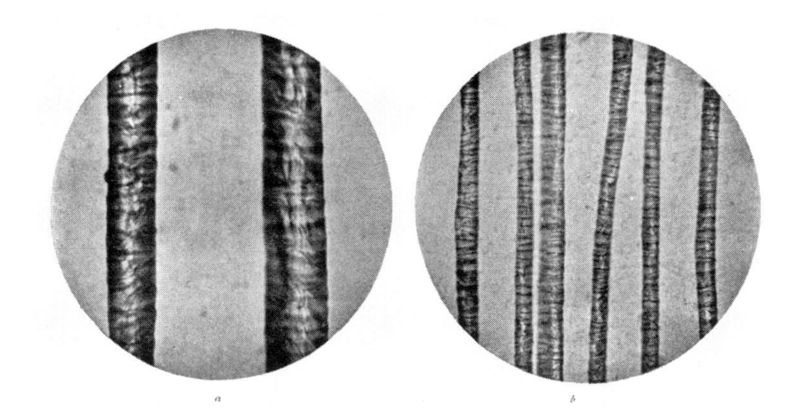

Fibres of (*a*) Cheviot and (*b*) Merino Wool. These images of wool fibres illustrate the highly-prized "superfine" qualities of Merino wool as compared to that of other breeds of sheep. This trait has been preserved for thousands of years, although it is of unknown origin, and may be the result of a genetic mutation. Image reproduced from J. H. Clapham, *The Woollen and Worsted Industries* (London: Methuen, 1907).

ewes crossed with wild North African stock from the Barbary Coast, were bred in Spain by the Roman agricultural revivalist Columella as early as the first century AD (Ryder 1983, 158, 250). In Spain, the breed was steadily improved throughout the Middle Ages with frequent importations of Barbary stock to become, by the seventeenth century, a distinctive fixed type, unlike any other breed of sheep in the fineness of its wool (Randall 1863). Whereas in most sheep the ratio of the thick primary wool fibres to the finer inner coat is approximately six to one, it is four times this in Merinos, at twenty-four to one. As Kenneth Ponting explains, the Merinos have the finest wool quality, or wool fineness: "In Britain, Australia and most parts of Europe, a system of numbers is used which relates to the size of worsted yarn that could be spun from that wool. The size of worsted yarn is measured in the number of hanks of 560 yards that weigh 1lb. A 60s (24 micron) quality wool is a wool that would produce sixty of these hanks (i.e. 60 x 560 yards) if spun to the limit of fineness" (Ponting 1980, 26). Merino wool is on average around twenty-one microns,[17] whereas the finest British wool, of the Ryeland type ("Lemster Ore"), for example, is 56s (twenty-eight microns), or 30 percent less fine than Merino wool.

By the fourteenth century, the great transhumances of Spain were well established, and the enormous annual migrations of Merino sheep and their shepherds protected under a monopoly known as the Mesta, a committee of noblemen, clergy, and wealthy landowners. The Merino sheep were organized into half a dozen great flocks, or cabanas, of which some became particularly well known internationally—such as the Negretti flock. The breed was of such importance to Spain that the export of a Merino sheep was a capital offence, and a privilege unique to the sovereign.[18]

With the rise of the bulk market in fine wools provided by Merinos, their importance to the British economy became more part of its colonial and imperial expansion than its domestic workings. Wool would never constitute as important a single component of the British economy once the Merino became the undisputed foreign source of superior fine wool in bulk, despite the fact that Britain remains the main producer of long wool breeds, such as Lincoln sheep. Within Britain, a return to the breeding of sheep as multipurpose animals, and the crossbreeding of distinct varieties to create a very large number of differently specialized breeds, reassumed a prominence in the eighteenth century that it has retained ever since. Today, the world sheep market still bears the traces of this seventeenth-century division, between sheep-breeding economies based almost entirely on Merino wool, and the British system, which relies on maximizing, stabilizing, and maintaining breed diversity, primarily for the purpose of mixing lines of sheep to diverse economic ends.

Border Crossings

In the same way the origins of the domestic sheep remain contested, so, too, does the precise definition of a breed remain somewhat arbitrary, at times closer to species, and at others simply a designation of variety or type (see Ritvo 1997). It is generally assumed that prior to the eighteenth century sheep were classified according to a range of strategies, with little or no written documentation of type or breeding programs.

At times, the main distinction among types of sheep appeared to be that between long-wool and fine-wool types, or simply between good wool-producing sheep and less good ones. At other times, it appeared that distinctive characteristics such as tails, horns, and shape made for the predominant classificatory features. Clearly there were early

In this eighteenth-century English engraving of unknown origin, sheep are designated by sex, country, region, and physical characteristics rather than by "breed" as it has come to be known. Historically, sheep were classified by their wool, type of tail, color, or origin in a particular region or flock before the advent of modern breed classifications.

regional and local classifications for sheep, which have been, since written records began, and are still considered to be very strongly shaped by their environment, including soil, climate, diet, and type of enclosure. It is likely the meaning of *breed* was somewhat more complicated in Britain, where a wide range of sheep types existed in correspondence with the widely varied topography of the British Isles.

Selective breeding has been practiced for millennia and constitutes one of the major means of domesticating animals from their wild type to become, among other things, white or hornless (polled). The main difference between a species and a breed could be defined in terms of breeding "out" and breeding "in": species are defined by reproductive isolation (the inability to breed out), whereas a breed is defined by reproductive consolidation (the establishment of type by breeding in), followed by reproductive concentration (fixing of type by in-and-in breeding). However, such distinctions are notoriously ambiguous and leaky. Long before modern breed lines or breeding methods were established, the carefully bred types of sheep created by the Romans or the Spanish cabanas might as well have been purebred in that they not only closely resembled each other but also passed on their characteristics, or bred to type. The careful selection of which tups would be used to work a group of ewes is well documented from the earliest written records, including the Old Testament and classical mythology, with advice to shepherds regarding the management of their flocks and the crossbreeding of different types of sheep, including specific trait associations such as that of colored wool with hardiness (Ryder 1983).

It is changes in the understanding of breeds, rather than what actually "constitutes" a breed, that are more readily identifiable, and with them changing ideas about inheritance. These are widely viewed to have undergone a substantial change in eighteenth-century Britain largely due to the work of the Leicestershire sheep owner Robert Bakewell, the first modern pastoralist to introduce a number of systematic refinements, in particular the use of in-and-in breeding, to alter the constitution of his animals in a dramatically short span of time.[19]

According to the early modern historian Harriet Ritvo, eighteenth-century attitudes toward animal husbandry favoured "the predominance of environmental factors such as climate and diet," with "no consensus about what could be inherited and how" (1995, 416; see

Directly descended from Robert Bakewell's revolutionary New Leicesters, the Border Leicester remains one of Britain's most recognisable breeds of sheep, and one of its most influential breeding stock. For the past 90 years all registered Border Leicester sheep have been tattooed in both ears to protect the breed's integrity. *Courtesy of the British Wool Marketing Board.*

also Russell 1986). Bakewell developed the sheep of his area, known as Leicester sheep, into what came to be known as New Leicesters, often considered the first modern breed of sheep. The old, "unimproved" Leicester sheep was a long-wool variety, long established and central to the long-wool trade, in which Britain remained dominant even after the rise of the Merino to world prominence. According to Trow-Smith (1959, 60), it is likely that the Leicester sheep was derived from the Lincoln or Lindsey types, in turn a legacy of Roman sheep farming. Robert Bakewell's New, or Dishley, Leicesters were bred to an ideal type, which was, consistent with the rapidly increasing demand for meat to feed expanding urban populations, a "butcher's sheep" that would provide "mutton for the masses" (Trow-Smith 1959, 61).

To produce the new barrel-shaped, short-legged meat animal out of the lanky and lean wool animal, Bakewell in-bred both dams and sires to establish his ideal type and fix the breed. Although it lost some of its wool-producing qualities, the New Leicester sheep fattened advantageously and could be brought to finish earlier, thus making it ideal for the mutton trade. According to the historians Roger Wood and Vitezslav Orel in their intriguing study of the importance of sheep breeding to modern genetics, Bakewell's sheep were considered revolutionary and astonishing, attracting international attention from visitors whose "amazement turned to incredulity as [they] learned of Bakewell's claim to have designed and reshaped this fast maturing animal in his mind before bringing it into existence. It was as though an engineer had invented a new machine" (Wood and Orel 2001, 95). Bakewell appeared to calculate his breeding strategy purely on the basis of appearance, adjusting it with crosses about which he was notoriously secretive. However, the New Leicesters bred true over many generations, enabling Bakewell to profit substantially from the high stud fees he charged other farmers.

As Ritvo explains,

> Bakewell claimed that when he sold one of his carefully bred animals, or, as in the case of stud fees, when he sold the reproductive powers of these animals, he was selling something much more specific, more predictable, and more efficacious than mere reproduction. In effect, he was selling *a template for the continued production of animals of a specific type*: that is, the distinction of his rams consisted not only in their constellation of personal virtues, but in their ability to pass this constellation down their family tree. (Ritvo 1995, 416; emphasis added)

This shift, Ritvo suggests, "represented the entry of a whole new course of value" into the livestock market—"a change in kind rather than (or as well as) a change in degree" (1995, 416). Developing the breed type as a breed line, which could be, in a sense, concentrated within a particular animal, both individualized the genetic capital and made stud services newly fungible as what might be termed breed wealth (see Franklin, 1997b).

As Trow-Smith notes in his account of Bakewell's livestock breeding innovations, Bakewell was able to transform the value of his flock through the intrepid sale and hire of his rams, which, between 1770 and 1789 saw an increase from five to thirty guineas for a season's use to as much as three thousand guineas—a hundredfold increase in fee for the same service in less than twenty years. The popularity of

Robert Bakewell inherited 440 acres from his father in 1760 and became one of the most successful livestock breeders in Britain whose influence continues to be felt. He introduced changes both to how animals were bred and to how they were marketed, which together transformed the livestock economy into its modern form. *Robert Bakewell* (ca. 1788–1790), John Boultbee. *Courtesy of the National Portrait Gallery, London.*

the New Leicester tup to introduce desirable qualities in other breeds of sheep was so great that, according to Trow-Smith, "there were only a few breeds in Britain into which some of 'Bakewell's legacy' was not introduced" (1959, 66).[20] This trend continued despite widespread evidence of its undesirable consequences, recorded by Trow-Smith, including the New Leicester's tendencies to obesity; loss of wool, hardiness, and milkiness; weakened procreative powers (due to the meat-shape "impeding parturition"); and "coarseness in the mutton" (so unpalatable "that no discriminating table would receive it") (1959, 65). In practice, rigorous culling was the only way to maximize the New Leicester's most desirable traits, namely, its capacities for rapid growth, desirable shape, and early finishing.

In sum, precisely the qualities that made the New Leicester useful for breeding in superior meat qualities could come at the cost of other valuable traits, recapitulating the age-old conundrum of improving an animal with so many uses. Moreover, Bakewell's methods are often referred to as the first truly modern or industrialized livestock breeding methods precisely because they are much more intensive, particularly in terms of the need for more aggressive culling. As Trow-Smith summarizes the dangers of Bakewell's innovative contribution to modern husbandry: "It is evident that, in capable hands, Bakewell's Dishley rams and the tups of other improved English Midland longwool flocks could add to other breeds that element of size and early maturity which the meat market now needed, without detriment to the constitution and the peculiar merits of the breeds upon which they were sparingly used. *Such work of improvement with so dangerous an instrument as the New Leicester called, however, for [a] delicate touch*" (1959, 69; emphasis added). In describing Bakewell as possessing a "husbandry genius" that called for sparing use of his "dangerous instrument" and a "delicate touch," as well as a readiness to cull any animals not fully conforming to his ideal type, Trow-Smith accurately portrays the mixture of talents and traits that "inspir[ed] in a multitude of his fellow stockmen the spirit of improvement" (1959, 69). Also evident in his assessment are the somewhat paradoxical aspects of the first truly modern sheep, such as the tension between the highly marketable shape and size of the animal carcass and the protection of other essential traits, such as fertility. A fast-growing sheep, ideally proportioned to become a terminal carcass was the profitable outcome of the New Leicester experiment. At the same time, the product and its means of reproduction remained somewhat at odds, as the advantages of larger size for breeding purposes were in direct conflict with those of the desired butcher-ready shape (which restricted the loins). The benefit of the rapid intensification of desired qualities, which was the main advantage of in-and-in breeding of parents and offspring to fix desired traits, consequently required a rigorous culling strategy. Thus the "delicate touch" required to achieve this fine balance required capable hands to prune, as well as to regenerate, New Leicester stock.

Breed Wealth

If, as Trow-Smith claims, the rationalization of land use in the late medieval period was the first truly industrial improvement of sheep breeding in Britain, followed several centuries later by the "modernization" of the animal itself in the eighteenth century, then in both cases the production of animal value through regional markets and international trade established a primary enabling condition of changes in sheep and sheep breeding. This integral and fundamental relationship between livestock, soil, and regional, national, and international economies was clearly perceived by Karl Marx in his description of the industrial revolution in the north of England as an example of the conditions necessary for the emergence of modern capital.

As Marx notes in volume 3 of *Capital*, "the process of Capitalist production as a whole" is directly based on former modes of production, and in particular agricultural production, such as tilling and cattle raising. He notes that "in the case of sheep-herding and cattle-raising, in general, as independent modes of production, exploitation of the soil is more or less common and extensive from the outset" (1972, 676). However, the confusion between modes of production and reproduction endemic to much Marxist scholarship on the economy is evident in his convoluted claim that "in accordance with the natural laws of field husbandry, capital—used here, at the same time, in the sense of means of production already produced—becomes the decisive element in soil cultivation when cultivation has reached a certain level of development and the soil has been correspondingly exhausted" (1972, 676). By this, Marx is referring to the use of grazing animals, and in particular sheep, to replenish "exhausted" soil by restoring its fertility. The soil, in other words, is cultivated in the symbiotic process of raising animals as "independent lines of production," and what appears striking is the way that the elementary reproductive components of this process are described as means of production *already produced*. Capital comes to be decisive, according to Marx, in terms of scale:

> So long as the tilled area is small in comparison with the untilled, and so long as the soil strength has not been exhausted (and this is the case when cattle-raising and meat consumption prevail in the period before agriculture proper and plant nutrition have become dominant), the new developing mode of production is opposed to peasant production mainly in the extensiveness of the land being tilled for a capitalist, in other words, again in the extensive application of capital to larger areas of land. (1972, 676–77)

Agriculture proper, with its concern for plant nutrition and the ability to till more extensive areas of land (from which surplus value and profit can thus be extracted), makes of capital something based on the scale of application. The primary processes of extraction—of value from animals and soil, as well as their reproductive interdependence—are clearly separated out from capital proper, becoming almost passive, or "natural," conditions for its emergence.

Yet the very processes of production literally at the root of this account of the emergence of modern capital, here depicted as a shift from subsistence agriculture to the precapitalist stage of peasantry, clearly depend primarily on elementary modes of *reproduction*, including what Marx described as the "differential fertility" of the soil (1972, 650). Returning to the earlier situation of sheep breeding in Britain, in the first major industrialization described by Trow-Smith for the late medieval period, it is a rationalization of land use, albeit facilitated by the large scale of manorial forms owned by Cistercian monks, which is seen to enable an unprecedented scale of sheep production, greatly increasing wool output for the global market.

Later, in the period of modernization widely attributed to the influence of Robert Bakewell, sheep are produced for a domestic economy, as meat for the growing proletariat class, who have moved off the land into large urban centers, or to work in semirural textile mills, and are producing goods for a rapidly expanding global trade.[21] Significantly, though, sheep are also acquiring the kind of value Ritvo (1995) describes as "genetic capital," as individual animals are seen to become repositories of what might be described as breed wealth. The outcome of this modernization is not only a new kind of wealth-accumulation strategy (the inflation of stud fees) but indeed a new kind of reproductive wealth: breed wealth (Franklin 1997a).

Putting together the elements of these transformations resituates control of reproduction—that is, the reproduction of soil and animals, as well as *their fertility*—centrally in the analysis of industrialization, modernization, or capitalization. This vantage point may be of particular importance for understanding the new industries of biotechnology, in which control over reproduction forms the basis of new productive processes through which the interests of nation and capital are united.

Enclosure

Control over land, soil quality, and livestock is not only acquired through rationalization and improvement but through crude forms of appropriation, to which sheep have proven instrumental. In Britain, the most dramatic and explicit example of the use of sheep to appropriate land is the enclosure of the Scottish Highlands in the late eighteenth century, a displacement of approximately fifteen thousand inhabitants to make way for profitable sheep-farming by wealthy landowning aristocrats.

The Scottish Highlands are renowned for their harsh and unforgiving climate and their exposed and rugged landscapes. The native inhabitants of the Highlands were, until the end of the eighteenth century, organized as clans headed by chiefs. They did not own land as property, but rather lived and worked on shared subtenancies leased to them by landed estate owners. Cattle raising brought a modest income to the estates and the Highlanders, but the local sheep populations were for domestic use, and consisted of small, primitive, and unimproved breeds.

A combination of economic pressure on large Scottish estates, often owned by English gentry, and the rapidly increasing demand for meat to feed Britain's expanding industrial workforce led, by the mid-eighteenth century, to the first experiments in large-scale sheep breeding in the Highlands. In 1762, John Lockhart-Ross of Balnagowan brought the first commercial sheep flock, of Blackfaced Lintons, into the Highlands region. They thrived well, in contrast to expectations, but were quickly succeeded by an even hardier breed, the Cheviot. So well bred, hardy, and productive a sheep had not been seen before, and the Cheviot was described by John Naismyth in a report for the Society for the Improvement of British Wool as "almost man-made," so ideally was it suited to the challenging climates of the north (qtd. in Prebble 1963, 25). The success of the Cheviot was confirmed in 1791 in the so-called Langwell experiment by the prominent Scottish agriculturalist Sir John Sinclair of Ulbster, himself a Highlander. According to John Prebble, in his vivid account of the Highland clearances, "Agricultural Sir John," as he described Sinclair, "was probably the only Scot of his age who used the word 'Improvement' objectively. Had he been listened to, had his example been copied, the half-century of evictions, burnings, riots and exile that followed

Formerly known as the Long hill sheep from the Cheviot hills, the Cheviot is the result of cross-breeding with a variety of larger and woollier breeds to produce a sturdy, tough, and profitable animal that by the late 1790s had become the so-called four-footed clansman of the Highlands. *Courtesy of the British Wool Marketing Board.*

might have been avoided. It was the kindly old man's tragedy that he brought the Great Sheep north for the benefit of his people, but was unable to prevent others from using it to oust theirs" (1963, 26). Sinclair brought five hundred ewes and a proportionate number of rams to his farm in Langwell in 1791, discovering, to the astonishment of his neighbors, that every one of his flock survived, and again in the year after. Long considered a debilitated and weak animal, the sheep, it was assumed, had to be housed in winter, was prone to diseases, and could not weather the extreme conditions of exposed mountain terrain in which even deer would not winter.

The surprising vigor of Sinclair's Cheviots set off an irreversible chain reaction. According to Prebble, the British Wool Society offered flocks of Cheviots to all Highland lairds "who aspire to the character of being active and intelligent improvers" (1963, 27). In spite of Sinclair's pleas for a gradual introduction of the animals, an "invasion" of sheep flooded into the Highlands. Having demonstrated that sheep farming with Cheviots could quadruple the profit from livestock per annum, Sinclair could only lament the results. According to Prebble, "So began the invasion of the Cheviot or True Mountain breed. They came up the old cattle roads into Argyll, Inverness, and Ross. They climbed up where the deer died, they throve where black cattle starved. Land which had produced 2d. (two pence) and acre under cattle now yielded 2S. (two shillings) under sheep. Four shepherds, their dogs and three thousand sheep now occupied land that had once supported five townships" (1963, 28). Under the banner of "Improvement," the "Year of the Sheep" led to riots and uprisings, followed by arrests and exile for many Highlanders displaced by "four-footed clansmen" (Prebble 1963, 25). Loss of land led to mass emigration from the Highlands,[22] followed by the forced clearances that began in 1800 on the estate of Lord Stafford (formerly the Most Noble George Granville Leveson-Gower and later the first Duke of Sutherland) who by marrying one of Britain's wealthiest heiresses, Lady Sutherland, had become the wealthiest landowner in Britain. As Prebble describes him,

> He was the Great Improver. Where there had been nothing in his opinion but wilderness and savagery, he built, or had built for him by the Government, thirty four bridges and four hundred and fifty miles of road. The glens emptied by his commissioners, factors, law-agents and ground officers (with the prompt assistance of the police and soldiers when necessary), were let or leased to Lowlanders who grazed 200,000 True Mountain sheep upon them and sheared 415,000 pounds of wool every year. He pulled the shire of Sutherland out of the past for the trifling sum of two-thirds of one year's income. And because he was an Englishman, and spoke no Gaelic, he did not hear the bitter protests from the poets among his people. (1963, 49–50)

Possessed of a "rage for Improvement," Stafford began forced clearances on his estate in 1807, legally evicting his tenants at will and then burning their homes, forcing them to flee leaving their crops, animals, and possessions behind. Many emigrated to the colonies, while others were offered uninhabitable coastal property where it was imagined they might earn a living from fishing. Patrick Sellar, one of Stafford's estate managers, described himself on his arrival in Sunderland as "at once a convert to the principle now almost universally acted upon in the Highlands of Scotland, viz., that the people should be employed in securing the natural riches of the sea-coast, that the mildew of the interior should be allowed to fall upon grass and not upon corn, and that several hundred miles of alpine plants flourishing in these districts in curious succession at all seasons, should be converted into wool and mutton for the English manufacture" (qtd. in Prebble 1963, 63).

Describing Sellar, Prebble claims that although "he would not have accepted the term, he was as much a colonist as those of his contemporaries who were preparing to dispossess the aboriginals of America, Africa and Australia to make room for meat, hide, and wool on the hoof" (1963, 63). The observation is especially apt given the importance of the Australian Merino sheep industry to Britain's ongoing economic dependence on the wool trade. The ethos of improvement that saw the Highlanders forcibly banished from their ancestral homes, and often left to starve, was based on enclosure of land as private property and the use of livestock animals to increase its profitability. This was justified in terms of progress, humanity, and even nature. Like the native inhabitants of Britain's colonies, the Highlanders were depicted as primitive, savage, and illiterate tribal peoples in need of rescue and improvement. Their land was unproductive, their way of life ancient and irrational, and they were being saved from themselves by being forced to leave their harsh way of life behind, along with their squalid homes. Ironically, it was more "natural" for sheep to inhabit the Highlands than the Highlanders themselves because the sheep made better use of it, in the sense that they were more efficient—meaning simply that they emerged as more economically profitable.

The pattern of clearances in the Highlands thus set an important precedent for the colonization of Britain's overseas possessions, in particular Australia, where an invasion of white settlers was facilitated by a swelling tide of white sheep. The pattern of pastoral land occupation to displace indigenous peoples, originally in New South Wales, Tasmania, and later throughout Australia, was accompanied by an ethos of improvement based on the twin principles of progress and economic gain. The imperial Roman wool trade, which first brought sheep in large numbers to Britain when it was itself an occupied Roman territory, was thus superseded in the eighteenth century by the promotion of sheep breeding in the British colonies as a means both of physically occupying them and integrating them economically into a system geared to increasing Britain's economic strength.[23]

Karl Marx was among those who protested the enclosure of the Scottish Highlands and the displacement of its inhabitants. Writing in the *People's Paper* in March of 1853, he denounced the expropriation of land from the ancient clans to create vast "sheep-walks." Explaining how the Countess of Sutherland had acquired her lands, "more extensive than many French Departments or small German Principalities," as a family inheritance, Marx denounced the injustice of her violent depopulation of them in favor of "economic reform":

> My lady Countess resolved upon a radical economic reform, and determined upon transforming the whole tract of country into sheep-walks. From 1814 to 1820, 15,000 inhabitants, about 3000 families, were systematically expelled and exterminated. All their villages were demolished and burned down, and all their fields converted into pasturage. British soldiers were commanded for this execution, and came to blows with the natives. An old woman refusing to quit her hut was burned in the flames of it. Thus my lady Countess appropriated to herself 794,000 acres of land, which from time immemorial had belonged to the clan. In the exuberance of her generosity she allotted to the expelled natives about 6000 acres—two acres per family. These 6000 acres had been lying waste until then, and brought no revenue to the proprietors. The Countess was generous enough to sell the acre at 2s 6d on an average, to the clan men who for centuries past had shed their blood for her family. The whole of the unrightfully appropriated clan-land she divided into 29 large sheep farms, each of them inhabited by a single family, mostly English farm-labourers; and in 1821 the 15,000 Gaels had been superseded by 131,000 sheep. . . . The British aristocracy, who have everywhere superseded man by bullocks and sheep, will, in a future not so distant, be superseded in turn, by these useful animals. (1853)

In describing the clearances as the unrightful appropriation of clan land, and referring to the systematic expulsion and extermination of the Highlanders, Marx echoed the protests of many critics who saw them as unjust and criminal acts of appropriation. In the view of Marx and many of his progressive contemporaries, the term *slavery* was not inappropriate to describe the treatment of the Highlanders, who were, in effect, stripped of their identity and patrimony as much as of their land and possessions. The depth of the disenfranchisement was compounded, in his view, by the fact of the Highland regiments having served the British aristocracy for centuries, both domestically and abroad, to secure possessions for the British Empire, which service was repaid by transforming their homeland into sheep walks.

Marx's somewhat ambiguous prediction that "having everywhere superseded man by bullocks and sheep" the British aristocracy would itself, in turn, be superseded by "these useful animals" remains unfulfilled, but it has suggestive implications. Writing retrospectively about the colonization of Australia—and the disappointed hopes of introducing sheep, and thus economic development, across its vast red deserts—the controversial antipodean historian Geoffrey Blainey has argued that the "elusive" truth that "in the harshest stretches of the continent the Aborigines were its quiet masters" was "learned slowly" (2001, 3–4) by its more recent occupiers. Decrying the white settler ethos of improvement, he goes so far as to commend Aboriginal peoples' protection of their lands against shortsighted government policies of development, arguing that such resistance may have done more to legitimate "Australia's possession of this empty and long-defiant territory" than "the constant eagerness to try new technology," which has had such "disastrous effects especially on fragile environments" (2001, 30–31).

Writing with similar concerns in 1792, the Year of the Sheep, the commander in chief of the King's Armies in Scotland, Lord Adam Gordon, wrote in a private letter to Henry Dundas, the home secretary, that

> everybody knows the wonderful attachment a Highlander has to his native spot, be it ever so bare, and ever so mountainous, and if these speculative gentlemen shall by any means, or from avarice, once dispeople their estates and stock them with sheep and that a bad season or two should follow, and the sheep be thereby destroyed, I am convinced no temptation under the sun will be able to bring inhabitants to such Highland property from any part of this world. (qtd. in Prebble 1963, 47)

Sheep Walks

Sheep have not disappeared from the Scottish Highlands; they remain plentiful across its otherwise deserted peat bogs and rugged, windswept countryside. The west coast of Scotland has become one of Britain's major tourist destinations, and is described in the January 2000 edition of the *Rough Guide to Scotland* as "the epitome of Bonnie Scotland" (2000, 507). It is, the guidebook's authors continue, "also the least populated part of Britain, with just two small towns, and yawning tracts of moorland and desolate peat bog between crofting settlements." Of the region's history, visitors are informed that

> The Vikings, who ruled the region in the ninth century, called it the "South Land," from which the modern district of Sutherland takes its name. After Culloden,[24] the Clearances emptied most of the inland glens of the far north, however, and left the population clinging to the coastline, where a herring-fishing industry developed. Today, tourism, crofting, fishing and salmon farming are the mainstays of the local economy, supplemented by EU construction grants and subsidies for the sheep you'll encounter everywhere. (2000, 507–8)

If the Duke of Sunderland's legacy of sheep bleatingly persists, his efforts to be remembered as a kind and benevolent laird have been less ruggedly enduring. The one hundred-foot-high Sutherland Monument, erected in 1834 atop the summit of Beinn a'Bhragaidh, near Golspie, is dedicated by "a mourning and grateful tenantry" to a "judicious, kind, and liberal landlord . . . [who would] open his hands to the distress of the widow, the sick, and the traveller." Living up to its name, the author of the *Rough Guide* notes that "unsurprisingly, there's no reference to the fact that the Duke, widely regarded as Scotland's own Joseph Stalin, forcibly evicted 15,000 crofters from his million acre estate" and that "campaigners are lobbying . . . to have the monument broken into pieces and scattered over the hillside, so that visitors can walk over the remains" (2000, 548).

Without doubt, the many legacies of sheep breeding are more fully "treaded in" in some parts of Britain than in others. In the wealthy home counties, London, and the south, sheep are barely visible in comparison to their substantial presence in the north of England and in Scotland. As the foot and mouth crisis of 2000–1, the subject of chapter 5, demonstrated, sheep remain a dominant feature of the British landscape, as well as the British economy, continuing to shape its trade relationships with former colonies. Thus sheep continue to be seen as a common thread uniting diverse features of contemporary British life, from its farming industry to its scientific accomplishments and its artistic creativity.

What is paradoxical about the way in which sheep can represent tradition and radicalism, conformity and creativity, or innocence and brutality, is similarly embodied in Dolly's simultaneous ordinariness and exceptionalism. A perfectly normal sheep, she is also a radically novel life-form. A picture of innocence, she is at once a reassuringly simple ewe and a threateningly unfamiliar clone.

Sheep do not acquire such a complex polysemic significance out of thin air, but rather through dense historical accumulation. This history is built in to their very conformation, both individually and as a population. As sheep have been shaped by the lands and soils they inhabit, so, too, have they been instrumental in altering the earth they treaded into arable soil, the salt marshes they transformed into mutton, and the high mountain heather they metabolized into wool. Thinking of sheep in this way, as bioreactors or forms of biocapital, enables their genealogies not only to be interpreted in terms of the lifeblood of a nation's economy but as conduits for breeding wealth. The meaning of the livestock industry, which has taken on a new economic significance in what Ian Wilmut calls "the age of biological control," can thus be situated within a much longer history of controlled reproduction—now extending to a domestication of the genomes of humans, plants, animals, worms, and microorganisms.

This view also allows a different perspective on the way in which the sheep-human relationship can be seen as genealogical: through it modern industrialized agriculture is linked both to the earliest pastoral economies and to cutting-edge bioscience via numerous vital or generative linkages, such as selective breeding. Sheep occupy a privileged place in human sociality that has, in some senses, been accelerated in the contemporary era through techniques that directly link the domestication of the human interior (the mapping of the human genome) to the redesign of livestock. In the same way that many British roads are built along the paths of sheep droves, the genomic pathways that will reshape the future of human health are being test-driven in the bodies of sheep. In this way, as in others, sheep substitute for humans in a subordinate and sacrificial role that is as biblical as it is contemporary, and as quotidian as it is revolutionary.

From the perspective of these visceral, institutional, and embodied legacies, Dolly must be seen as a product of the same agricultural, industrial, and economic inheritance that has given Britain its constitution as a nation. She epitomizes the combination of agricultural,

The Scottish Blackface sheep, or "Blackies," are the most numerous sheep in Britain. A hardy mountain breed, they have remarkable survival abilities, often attributed to their detailed knowledge of the hills onto which they are "hefted," and to which they regularly return each winter. Dolly was gestated by two Blackface ewes. *Courtesy of the British Wool Marketing Board.*

scientific, and industrial innovation that has made Britain both a leader in stem cell research and in sheep breeding. She is a brilliant, but practical, innovation, designed to increase scientific knowledge, improve human health, and create new markets for commercial products. In all of these respects, she demonstrates the enormous scale and complexity of control over the means of reproduction, much as Marx described in his accounts of the integration of seeding land, feeding animals, accelerating propagation with manure, and reclaiming nonarable land through the use of sheep. She, like her parent institute Roslin, is a viable offspring of Britain's emergence as a nation and as an empire, as well as being a poster sheep for British biotechnology in the competitive climate of the knowledge economy within Europe and globally. Her ultradomesticated substance, which she shares with the many sheep who did not, unlike her, achieve viability in the experiment that led to her creation, stands as a reminder of the costs of improvement and scientific progress in the realm of livestock breeding. As a national animal, and a symbol of British innovation, Dolly thus poses an ongoing question about the consequences of reengineering the germplasm in the pursuit of better breeds.

In turn, the new frontier of Dolly's genomic interiority confirms the close relationship between domestication and unexpected effects, much as in the history of livestock breeding an intensification of methods has brought with it new risks of introduced pathology. As in Marx's discussions of soil, the regulation of fertility is as important as proof of viability in the effort to create sustainable and profitable "lines," be they of pure bred sheep, immortalised human cells, or marketable agricultural products. Without adequate control over the means of reproduction, and a suitably complex understanding of their structure and function, productivity may decline or even cease. Dolly is thus an animal who brings us to a turning point in the history of agriculture, medicine, and science, because her viability bridges the before and after of a revolutionary means of propagation in many ways comparable to the experiments of Robert Bakewell, and because she signals the importance of reproductivity as more than a background, "natural," or passive mechanism. Like Bakewell's Dishley sheep, Dolly is the animal model for a technique that has become one of the basic elements in the production of new markets and products that belong to a new form of capital, biocapital, and a new form of harnessing cellular vitality as totipotency, immortality, transdifferentiation, and regeneration. As such she is also a very British sheep who shared her constitution with the soil on which she was born and bred every bit as much as the scientific research facility in which she was designed and made.

Colony

It is not unusual in the history of organisms for the protoplasm to remain in a state of retarded development—almost quiescence—for some considerable time and then suddenly to commence growth at practically a forced rate, as if to make up for the delay previously experienced. So it was with the early history of Australia—so much so that the observer may point to a single year and say that then commenced the real story of Australia's progress. This point, when the graph suddenly took its upwards curve, was the period 1835—when the protoplasm began to indicate what form of organism it was developing into, when the infant community began to assume its distinguishing characteristics.
—Stephen Roberts, *The Squatting Age in Australia, 1835-1847*

The germ of the distinctive "outback" ethos was not simply the result of climatic or economic conditions. . . . It sprang rather from their struggle to assimilate themselves and their mores to the strange environment.
—Russell Ward, *The Australian Legend*

As everyone knows, sheep came to Australia with the first settlers and any history of the sheep industry is in fact a history of white exploration and expansion across the continent. In the 1950s we were even said to be riding on the sheep's back, although the most valuable commodity then was the sheep's wool, not its meat.
—Stephanie Alexander, *The Cook's Companion*

4

Sheep are often considered to be synonymous with the founding of Britain's last, and largest antipodean colonial possession, *Terra Australis*. However, the original colony of New South Wales was not originally imagined in a pastoral idiom. When the First Fleet touched ground at Cape Cove in 1787, the British ("Home") government's intentions for its new possession were strictly import only: it was in-

tended that the remote colony become a jail in order to relieve Britain's overcrowded prisons of a significant portion of their convict population.[1] To this same financial end it was intended that the colony should become economically self-sufficient. As the historian G. J. Abbott claims in his account of early Australian settler-colonial history, "Whatever ultimate reasons might have been implied in the British government's decision in 1786 to found a settlement on the east coast of Australia, the immediate explicit aim was to establish an economical prison . . . that by the end of the second year of its existence . . . would be self-sufficient" (1971, 17). The first sheep to set hoof upon Australian soil, the ninety or so on board the First Fleet who survived the passage, were thus intended for subsistence only and were so poorly cared for that within a year all but a few of their number had perished. In 1793, 110 additional sheep were unloaded in Sydney from the Indian ship *Shah Mormuzear*, bringing the total to 516 head in July 1794, when the first official livestock count was taken (Abbott 1971, 23).[2]

Most of these "first sheep" were owned by English civil and military officers—members of the most privileged sector of early colonial New South Wales society who later became its private pastoralists and privileged elites, such as Captain John Macarthur. These often shrewdly entrepreneurial livestock owners were generously rewarded by both the home and colonial governments for increasing the size of their flocks in order to reach the desired state of economic self-sufficiency as rapidly as possible. This goal, initially at odds with fine-wool production, nonetheless aided its early development by discouraging the use of livestock for meat (in order that the animals be bred as much as possible) and by, essentially, privatizing sheep raising among colonial elites eager to rank among the first to take advantage of export opportunities—and having the means to do so.

New South Wales has been described as having an ideal climate for the production of fine-wooled sheep, but sheep are known to thrive in a more diverse set of landscapes than almost any other animal. Indeed, few animals other than dogs exist in so varied and numerous a range of global habitats. Even Merino sheep, despite their African and Asiatic origins and largely Mediterranean cultivation until the 1800s, have successfully been introduced to the harshest environments, such as Iceland, where they winter as happily as seals. They are equally robust in the face of drought, heat, dust, and the extremes of weather and disease imposed by a tropical, or desert, climate, and can even be raised indoors.[3]

Sheep were useful in colonial Australia for many of the same reasons they have been a productive animal elsewhere. Requiring little to no maintenance from their keepers, and being highly mobile and tractable, they provided a reliable and efficient source of milk, wool, meat, lanolin, and other products including hide, manure, fleece, and tallow. Capable of living where other animals will perish, in part because they graze more efficiently than any other ruminant, and of doing without water for prolonged periods, sheep proved the ideal settler animal for colonial Australians, providing cheap subsistence on the hoof. Once the Blue Mountains had been crossed in 1813, the spread of settlement rapidly followed the sheep runs out into the rich pasturelands to the West. Indeed, the settlement of New South Wales, often referred to as Australia's first frontier, was largely a settlement by sheep.

This spread of settlement by sheep had consequences for the indigenous, non-landowning, and subsistence economy-based people of what came to be known as Australia not entirely dissimilar to those faced by the inhabitants of Britain's last frontier in the Scottish Highlands during a slightly earlier period (Cheviots were brought to Langwell in 1791). As Henry Reynolds describes in his account of Aboriginal reactions to white settlement, pastoralism posed an enormous challenge to indigenous communities, "altering ecologies and disrupting traditional economies" (1990, 159) to the point of making them all but uninhabitable by Aboriginal people unless they adapted to, or joined, the settler way of life. Not surprisingly, indigenous peoples resisted this pastoral invasion. In his chapter entitled "Black Shepherds," Reynolds notes that many Aborigines included sophisticated herding techniques among their hunting skills, such as the production of long races of sticks, bows, and bushes into which marsupials and emus were mustered "like flocks of sheep" (1990, 159). Using horses and dogs captured from white settlers, Aboriginal hunters regularly cut out whole sections of cattle herds and sheep flocks for their own use, sometimes yarding them for long periods hidden in the bush.

As the anthropologist Patrick Wolfe writes: "Sheep competed with indigenous fauna for subsistence, consuming the tubers, shoots, and seeds whereby the indigenous fauna reproduced itself and rapidly re-

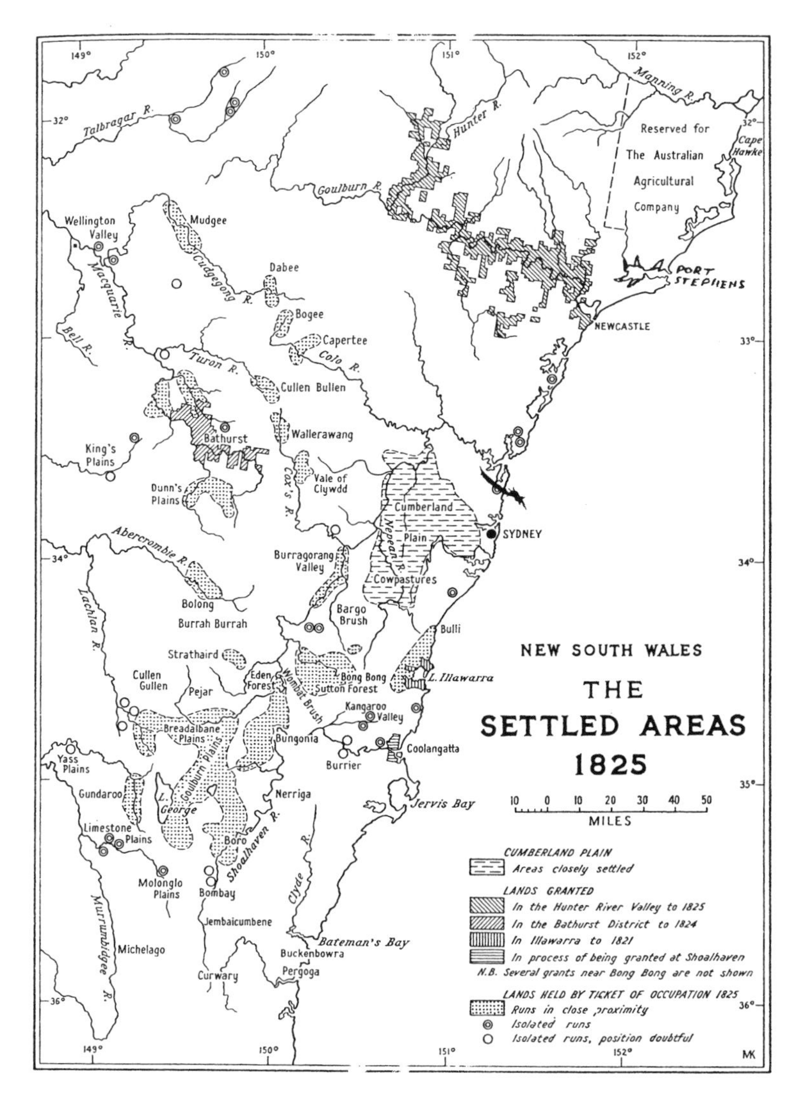

Australia's first frontier, to the west of Sydney, was largely a settlement by sheep, and the occupation of land was largely by a white tide of Merino sheep sweeping westward from Sydney to the Cumberland plain. *Reprinted from Charles T. Burfitt,* History of the Founding of the Wool Industry of Australia *(Sydney: William Applegate Gullick, 1913).*

ducing waterholes to mud. In a short time the only subsistence remaining available to indigenous humans was the introduced fauna whose protection was axiomatic to the pastoral project" (1994a, 184). And, as Frederick Rose similarly confirms, "Undoubtedly, the advent of the white man in 1788, particularly with the introduction of sheep-raising on an extensive scale after 1820, had a most devastating effect on the Aborigines, leading to their virtual annihilation in the settled areas" (1965, 15). Thus, bred white for the European fine-wool trade and treading themselves in across New South Wales, Tasmania, Victoria, Queensland, and Western and South Australia to become part of the landscape, sheep were essential vectors of Australian colonization. While brokerage houses in faraway London annually determined the market value for their wool, providing Australia's first significant export commodities, the quotidian, everyday use value of sheep at home contributed to the settler economy directly. In effect, sheep converted pasturage into profit, and then moved on. A mobile white sea of lucrative wool—with additional spin-offs in the form of milk, cheese, mutton, and fleece—Australian sheep were used not only to displace indigenous people but to ensure the nonreproducibility of the subsistence ecology supporting the Aboriginal way of life, thus literally entrenching the settler economy into what was to become their national soil.

Thus, although the "white sails" of the first European ships to arrive in Australia are often figured as the harbingers of dispossession for the land's indigenous inhabitants, it was the white tide of sheep soon outnumbering humans in the antipodes who constituted its terrestrial modus operandi. Sheep consequently pose specific questions for the understanding of colonial occupation and its organization that recompose colonialism's agents and agencies and provide a means of reconsidering how Australian national genealogies, sheep breeding, and the making of the Australian "frontier" are intertwined.

This chapter explores the significance of selectively bred sheep to Australia's settler economy both in order to consider the social, ecological, and geographical consequences of its four-footed frontier and to examine how Australia's frontier legacy in turn becomes genealogical by establishing the origins of its national identity. This frontier effect—whereby it becomes, in the words of Patrick Wolfe, something that is "carried in the veins" of contemporary Australians (1994a, 1994b, 1998), bears a close resemblance to the idea of the heritable frontier identity introduced in Frederick Jackson Turner's famous frontier thesis concerning the process of Americanization. These genealogical aspects of frontier heritage and national identity, which in Australia have a distinctively sheepish cast, are further complicated by the emergence of new national biological frontiers in the forms of stem cells, genomics, and reproductive biomedicine, which are literally remaking of the human interior a new cartography of both vitality and economics. Significantly, these, too, reveal an ovine descent pattern as many of the components of IVF, and in particular ovulation induction, were perfected in sheep. Thus, moving on from the previous chapter, in which the thickness of Dolly's genealogy was linked to her importance as a national animal, this chapter locates her within Britain's wider colonial heritage, through which experimental sheep breeding, and the bioeconomics of the wool trade, can be seen to have laid vital scientific, commercial, and agricultural foundations for contemporary partnerships between Britain and her commonwealth partners in the pursuit of future bio-innovation.

Wool to Wealth

Much dispute predictably surrounds the origins of the fine-wool trade in Australia, and the role of certain prominent Australian "forefathers" within it, but most commentators identify a significant change in the early colonial economy in the period 1803–4, under Governor Philip King, when the colony began to reach self-sufficiency, and the possibility of export, or staple, commodities arose in earnest. Prominent among such export opportunities was fine wool, a product that had already begun to be advanced through selective breeding programs employing imported Merino sheep by private pastoralists including the Macarthur, Marsden, and Riley families (Abbott 1971, 17–47). According to Jill Ker Conway, one of the first Australian historians to research the early sheep-and-wool economy in detail, the fine-wool industry of New South Wales developed unevenly from 1803 to 1835, beset by a number of compromising factors and delayed from achieving significant expansion until 1820. These factors included the lack of proper breeding stock or knowledge of breeding, the small size of the labor force, the absence of established markets for wool, and the conspicuous lack of support from the British government for fine-wool production, evidenced by, among other restrictions, Governor Lachlan Macquarie's failure to facilitate livestock pasturage in the rich valleys made available in 1813 by the successful crossing of the Blue Mountains (Ker 1960, 1961, 1962). Other historians including G. J. Abbott (1971) posit different factors and disagree about the attribution of delay to the wool trade, while concurring with Ker Conway that it only became fully viable after more than two decades of fits and starts.[4] According to Abbott, "1822 was when the accelerator was pushed to the floor," leading to the wool boom period of 1822–51, widely described as Australia's "Great Pastoral Age."[5]

The beginning of the pastoral age, according to the celebrated Australian historian Stephen Roberts, whose aptly genealogical protoplasm analogy is reproduced as the epigraph to this chapter, was "the period 1835," when "the early history of Australia" and the "real story of [her] progress" began "suddenly to commence growth at practically a forced rate" (1964, 1).[6] Roberts's protoplasm analogy is, as Abbott points out, "an embellished form of [the historian James] Collier's [1911] contention that *the legalization of squatting* [proposed 1835, enacted 1836] represented 'the true germinal protoplasm of the British colonies at the Antipodes' " in his classic account of *The Pastoral Age in Australasia* (Collier qtd. in Abbott 1971, 4, emphasis added).

In an analogy that complements more recent sheep breeders' successes in the arts of propagation, Collier suggested not only that Australian history was essentially pastoral, but that "the germinal protoplasm of pastoralism stretches from age to age and spreads from this country to that, but is always and everywhere self-identical. It is an undying chain, which began with the first domesticated herd of cattle or horses, the first tamed flock of goats or sheep, is still vital, indeed is more vigorous than ever, and will live on till mankind returns, at the end of its long parabola, to a condition resembling its primitive state" (Collier 1911, 7). Collier was not alone in proposing biblical righteousness to the "real life of Australia," driven by "the overflow of unauthorized squatting beyond the boundaries" and unsuccessfully staunched by the British authorities. As he notes, the Home (British) government urged its colonial authorities to "put down such unauthorized squatters," but it was "too late" (Collier 1911, 4). In vivid language he depicts the crucial "passage in Australian history" that quickened into life "the true germinal protoplasm of the British colonies at the Antipodes, their substance and inner life . . . out of which all else has grown":

> The local Government was alarmed by those ugly ducklings, who thus boldly took to the water. Urged by the Home Government to put down such unauthorized squatters, it endeavoured [*sic*] to suppress them. It had to be determined whether those immigrating pastoralists should be hunted like wild animals and turned into Bedouins of the desert, or should be encouraged to grow up like the patriarchs of old, nursing within themselves the germs of the future State. The English authorities hesitated long, seeing their plans for Australia on the point of being completely overthrown by events they never could have foreseen. At first, the Imperial Government decided in favour of adhering to their original design, and instructions were issued to arrest the overflow of unauthorized squatting beyond the boundaries. It was too late; the colonial Canute who would have said to the waves: "thus far, and no further," would have been as impotent as Mrs. Partington with her mop.[7] Two wise Governors endeavoured [*sic*] to regulate the inundation they could not dam, and by a series of ordinances they introduced law and order into the lawless doings of the adventurers. In these they saw the promise of opulence and of a mighty State. These sagacious men reported to that effect to the authorities in England, and succeeded in persuading them of the justice of their views. Squatting was legalized and regularized, and, giving an impetus to free colonization, it lifted the community to a higher plane, and started it on a new career. (Collier 1911, 4)

In this complex account of the "impetus to free colonization," which directly precedes Collier's germinal pastoro-plasm analogy, a mixture of justifications are provided for the "inundation" of illegal squatters into the Australian interior, driven by the sheep mania of the 1830s.

This flood is depicted as at once "legalized" and "set free," while also being a force of nature, as in a sea or tide that could not be controlled. A crescendo of similes elevates Collier's "immigrant pastoralists" from a state of nature ("wild animals") to nomadic barbarism ("Bedouins of the desert"), finally becoming the embodiment of civility ("like the patriarchs of old"). Collier's primordial pastoralists are thus the bridge between savagery (ugly, bold, and lawless), and civilization ("nursing within themselves the germs of the future

State"). Importantly, they are seen not only to have created "the real life of Australia" but to embody this generative potential, comprised in hybrid parts of their raw, untamed vitality and the natural rightness of their pastoral urges. It is thus a genealogy of pastoralism that establishes in them the rightness of both nature and civilization.[8]

In turn, as "the pastoral community was henceforth free to develop along its natural lines," Collier explains, "all else" Australian grew from this "root." Out of the "pastoral and central life of the Australian communities" rise up "its nurslings, like the mechanical industries, or spring up out of it by natural growth, as agriculture and horticulture" (1911, 5). As he concludes, "All this is the work of the Golden Fleece" (1911, 6).

The White Invasion

The work of the immigrant pastoralists in Australia may have proceeded along the straight boundary lines, parallel wagon tracks, and evenly ploughed furrows of regulated settlement, but these were not the natural lines of the societies who inhabited those regions prior to the arrival of the waves of white sheep and settlers who occupied their land and destroyed their way of life. In their rise from "wild animals" to "Bedouins" to become "patriarchs" and protostatesmen, the "immigrant pastoralists" underwent a transformation from "ugly" lawlessness to become the very embodiment of the Australian way of life, but this required that another lawless, ugly barbarian should fill the "savage slot."[9] Thus, as Patrick Wolfe argues (1994a, 1994b) a direct genealogical consequence of the frontier effect through which sheep and pastoralism "built" Australia was the binding together of white settler immigrants against a binary separation between black and white Australians that has persevered with noticeable tenacity.[10] This violent process of historical estrangement is in turn repeated in the flows connecting Britain to her colonies, not only because the white settler occupiers of New South Wales originally fled or were deported from England, Scotland, Ireland, and Wales but because it had often been these peoples' own forced eviction from their lands that had led them either into crime or a poverty from which emigration was one of the only avenues of escape. It is in this way that the transnational genealogies of colonial diaspora remain haunted by their own historical alienation.

The white tide of settlement that brought into being first the colony of New South Wales and later the nation-state of Australia thus illustrates several interconnected levels of what we might call the economic geography of postcolonial genealogy, in terms not only of what it can connect but what gets trimmed or pruned from its branches. Moreover, this work of genealogical connection and disconnection can be analyzed as both constitutive and performative in ways that are relevant both to the flows between colony and empire and the ways in which these come to be embodied. In turn, this creates a method of genealogical reckoning that not only looks back at how certain conditions came into existence but at how they are orientated toward the future, and how they can be seen as promissory repositories of purpose *toward* particular ends.[11] Since it is the toward (what future is she pointing toward?) as well as the from (how did she come into being?) of Dolly's body this book seeks to understand, these perspectives on genealogy are inevitably in need of rumination.

Frontier Legacies

In the foreword to *Caledonia Australis: Scottish Highlanders on the Frontier of Australia*, the historian Don Watson is quick to note the "inescapable ironies" of the process by which the displaced peoples of one of Britain's last northern frontiers became the settler-colonizers of its newest southern one. "Like [the Duke of] Cumberland they justified their destruction of an ancient way of life in the name of advancing civilization," Watson claims (1984, 217). Emphasizing the importance of the frontier not only as a place of settlement but also as a potent source of national mythology, Watson reminds his readers that its legacies inevitably constitute recapitulations of the crimes and injustices committed along its borders: "The end of a frontier era is marked by the emergence of an historical orthodoxy which cannot incorporate the losers. And when those 'passing' societies are examined they are found to be pre-eminently built on . . . myths" (1984, xi).

In Australia, as in New Zealand, the "emergence of an historical orthodoxy" underpinning the "passing" of traditional Aboriginal society, and the mythology established to legitimate the nation's emergence, are primarily based, as in the United States, on the significance of the frontier, which, like a curtain, is opened to reveal the heroic exploits of pioneering white settlers, and closed to obscure the brutality of their occupation. As Patrick Wolfe astutely characterizes the work of the frontier, the "primary paradigm" of Australian settler-colonization, "Empiricially, rather than being fixed, as in the visual metaphor of a dividing line, the 'frontier' was shifting, contextual, negotiated, moved in and out of, enacted and suspended. . . . In short, it is necessary to distinguish between the misleading or illusory nature of the concept of the frontier as a representation and the social effects that were sustained by the currency of that representation" (1994a, 95). "What matters," Wolfe continues, "is that [the frontier] was a *performative representation*" (1994a, 96; emphasis added)—the *idea* of the frontier was a formative concept because it could direct activity, shape settler-colonial perceptions of themselves and their conditions, and provide the legitimating context for a discourse of occupation as improvement.[12]

The brutality with which these supposed improvements were imposed often surpassed the alleged savagery it was their legitimating purpose to reform. Watson's chronicle of the settlement of Gippsland in southeast Australia, largely but not exclusively occupied by immigrant Highlanders, details the tragic and relentless extermination of native Aboriginal people—from the Warrigal Creek massacre to the spontaneous "black-hunts" pursued as a sport by settlers such as Patrick Coady Buckley, who waged his own personal war with members of the Kurnai people throughout the 1840s. As Buckley wrote in his diary on 18 January 1845, at the height of Australia's "Great Pastoral Age": "I saw two Blacks coming along the Beach from near the Creek [and] waited behind a sand hammock until they came opposite to me. I then rode towards them and they took to the sea. I had pistols with me and fired Blank Shots to keep them in the sea which I did for almost four hours and drove them along in the water . . . about a mile [until one of them] nearly drowned in the breakers" (qtd. in Watson 1984, 226). At the same time that illegal companies of armed men, such as the Highland Brigade, were organized to hunt down Aborigines in the bush,[13] immigrant settler communities became increasingly dependent on Aboriginal knowledge, skills, assistance, and unpaid labor in order to establish their way of life. As Watson recounts,

> White settlers used Aboriginal skills *to explore the district and to tame it*. Every settler had at least one "boy" as a stockman, tracker, and general labourer. McMillan,[14] Raymond and several other squatters put Aborigines to work washing sheep. When shepherds left for the gold fields, Aborigines took on their jobs. They were excellent stockmen. Buckley used

> them for all sorts of tasks. Other settlers, Tyers among them, employed Aborigines "even as house servants." Frederick Jones told Tyers in 1853 that without them he could have neither washed his sheep nor shepherded them; he could not have harvested his wheat crop, "nor carried on the ordinary work of the station." (1984, 237; emphasis added)

Although hunted like wild animals and exterminated as if they were vermin, the indigenous people who worked as stockmen, explorers, household servants, and agricultural laborers for the early white settlers had inhabited the Australian continent for over forty thousand years and were thus irreplaceable in the task of its "taming" to fit the needs of its new inhabitants.[15]

Beyond the Pale

A key transition in the history of the Australian frontier is the point at which its legacies come to be seen not only as shared and embodied as a bond uniting white settlers of otherwise disparate ancestry but as having created something among themselves not only palpable and heritable but distinctive and new. This "germ" of national identity is widely depicted within the nostalgic literature celebrating the transformation of "wild" outback culture into a "civilized" fraternity as having reached its tipping point, or "births" in the midst of the sheep mania accompanying the economic success of the early wool boom. Roberts's "great age of pastoral occupation" and Collier's "squatting age" are widely regarded as the source of the archetypal Australian character. As Roberts describes the birth of a new "emotional patriotism" in the 1830s,

> Nobody minded being called a "squatter" now. . . . The mania was in full swing, and every able-bodied man thirsted for the bush and pined to ride in the dust behind masses of smelling sheep and live on an unchanging diet of mutton chops, unleavened damper, and post-and-rail tea. It was something in men's blood, like the emotional patriotism of a war period or the unnatural stimulus of a gold rush. . . . The bush, the sheep, the clipper on the tide—the process almost ran like a refrain in men's minds, and the community sang their song of the western wagon and turned toward the interior. (1911, 9)

According to Roberts's vividly corporeal account of the archetypal squatter, he was a prototypic Australian consumed by a thirst for the

Merino flock in New South Wales. Australian pastoralism played a central role in both nation formation and land occupation during the period of colonial settlement. The prototypic figure of the squatter on the outback is an equally abject, liminal, and heroic character in the national mythology of colonial settlement. *Reprinted from J. H. Clapham,* The Woollen and Worsted Industries *(London: Methuen, 1907).*

bush who shunned the effeminate comforts of domesticity in pursuit of war, gold, sheep, and the interior.

Almost like sheep themselves, this "tide" of able-bodied men marched behind "the western wagon" fueled by the "emotional patriotism" that drove them "like a refrain" toward the edge of the frontier. This view remained virtually unchanged fifty years later in textbook accounts such as that of Russel Braddock Ward, whose postwar ode to the "noble bushman," *The Australian Legend,* first published in 1958, is all but required reading for every Australian schoolchild. Harking back to the germ theory of history once again, Ward's central claim was that it was not so much pastoralism itself, but the struggle to establish a pastoral way of life against the adverse "conditions" in a "strange environment" that made Australia: "The germ of the distinctive 'outback' ethos was not simply the result of climatic or economic conditions. . . . It sprang rather from their struggle to assimilate themselves and their mores to the strange environment" (Ward 1978, 33). Significantly, the frontier effect described by Ward borrows not only from the earlier pastoral odes of Roberts and Collier but from Frederick Jackson Turner, the American historian who proposed his famous frontier thesis of American cultural independence from Europe in 1813 (the same year the Blue Mountains were crossed to find more grazing land for sheep in New South Wales)—a model that became an influential lens for understanding Australian history in the postwar period.

Frontier Fathers

The word *frontier* comes from the French *frontière* meaning "to front" or "to face." It is the opposite of *derrière*, or "behind." Etymologically it has many senses, most of them recent—and military. To make a frontier against, *faire frontière*, is to defend or protect. A frontier is the part of one country that fronts or faces another. It is a border—be it of a far edge or a remote region, or the immediate edge of a fortress. The frontier is also defined in terms of settlement, as its margin. As a descriptor, *frontier* is associated with a crude and rudimentary or primitive way of life, as in *frontier mentality* or *frontier cuisine*.

In all of these senses, the frontier is a place, and indeed a border, limit, or by implication, a line.[16] Like a railway line, it has two sides, and like the lines of a playing field, it has an inside and an outside. Ironically, to be inside the frontier is to be facing the interior, which lies beyond it. This brings us to the work of the frontier not as a place but as an idea, a concept, or an idiom, as in *the frontiers of knowledge*, which in turn leads us also to the very important ideological work of the frontier, perhaps epitomized by its mythologized role in the formation of American culture, but equally important to the Australian national imaginary.

Frederick Jackson Turner, the U.S. historian hailed as the "Father of the Frontier" and the founder of the so-called frontier school of history, proposed in his famous thesis, "The Significance of the Frontier in American Society," that the frontier was a crucible that created a new American native out of the conditions it imposed, constructing the "sacred" frontier heritage of individualism and self-reliance and thus equating the frontier with a process, a condition, and an embodied legacy.[17] It was Turner who reimagined the national frontier as both generative and heritable, thus giving it significant ideological power as an imagined source of supposedly original or native American identity: according to his thesis, the frontier proved as essential and foundational to the character and values of any true-blooded American as it was productive of them to begin with. Importantly, however, Turner's "native American" was selectively bred: he was a white European settler crossed with indigenous influences. Consequently, neither European nor "Red Indian," Turner's frontier nativity scene focused on a new breed of white settler who had been "born again" in the conquest of the interior.

Turner's own genealogy, in both familial and scholarly terms, was similarly hybrid. A frontier child himself, and writing in explicit opposition to his Europeanist training in history at Johns Hopkins University in the germ theory of history exemplified earlier in the work of Collier (1911),[18] Turner introduced a new form of historical vitalism in his famous essay, first delivered as a lecture to the World's Congress of Historians in Chicago in 1893, which opens with an image of institutions and constitutional forms as organs and the frontier conditions as "the vital forces which call these organs into life" (1961, 37).[19]

Turner himself refers repeatedly to the frontier as a line and as a place. In his view, however, it was never simply an edge or a border, but something much more complex and commodious: its constant westward movement required a ceaseless "return to primitive conditions on a continually advancing frontier line, and a new development for that area." Consequently, as Turner put it, "American social development has been continually beginning over again on the frontier. This perennial rebirth, this fluidity of American life, this expansion westward with its new opportunities, its continuous touch with the simplicity of Indian society, furnish the forces dominating the American character" (1961, 38). For Turner, then, the frontier was many things: a place and a process, a line and a condition, a historically advancing border of settlement, a perennial rebirth along its borders, and afterward a legacy embodied as the American character. Significantly, his definition of the frontier, while distant, was in constant contact, and indeed even "continuous touch," with what it faced or fronted, making the borderlands of the frontier both a place of opposition between inside and outside, *and a place of their continuous mixing, interpenetration, and hybridity*.

This forge, or crucible, of the frontier is thus a site of generative mixtures, indeed of generativity through mixture, or contact, producing both new opportunities and new entities, such as "the American character." At once a border, or line, the frontier is also a space of constitutive relationalities, or ties—a "cord of union" (Turner 1961, 46).[20] The contact zone, as Mary Louise Pratt calls it in her analysis of the formative intimacies of imperialism (1992), is a place of close encounters.[21]

These intimate encounters are at once generative and destructive, tragic and romantic, engendering both hope and despair. The pioneer spirit associated with frontiering is paradoxically depicted as a union with, triumph over, and failure to defeat the obstacles it faces or fronts. It is this mixture of agencies and proximities, often combined with a lack of agents or scaling devices, that gives the idiom of the frontier its mythic proportions and ideological power.[22]

Turner refers in a remarkable passage to the "flow" of the frontier on historical maps, and to how the lines on these maps incorporate other lines, such as the paths used by Native American peoples, thus tying them together in a kind of palimpsest of "advance."

> Study the maps . . . in which the settled area is colored . . . and you will perceive that the dark portion flows forward like water on an uneven surface; here and there are tongues of settlement pushed out in advance, and corresponding projections of wilderness wedged into the advancing mass. The [next] map will show gaps filled in, and the process repeated on a new frontier line. . . . Investigation will reveal the fact that settlement has not only flowed around physical obstacles, following the lines of least resistance, but that the location of the Indian tribes has been influential in determining the lines and character of the advance. (Turner 1961, 31)

Shifting his emphasis from the lines of settlement to the lines of commerce, Turner's analogies become increasingly corporeal, fusing together the vitalities of bodies and the land, and culminating in a strikingly modern image of complexity in which the lines of aboriginal intercourse are interpenetrated by the lines of "civilization."

> The student of aboriginal conditions learns also that the buffalo trail became the Indian trail, that these lines were followed by the white hunter and the trader, that the trails widened into roads, the roads into turnpikes, and these in turn were transformed into railroads. The Baltimore and Ohio road is the descendant of the old national turnpike, and this

> of an Indian trail once followed by George Washington in his visits to the French. The trading posts reached by these trails were on the sites of Indian villages, which had been placed in positions suggested by nature, and these trading posts grew into cities. Thus civilization in America has followed the arteries made by geology, pouring an ever richer tide through them, until at last the slender paths of aboriginal intercourse have been broadened and interwoven into the complex mazes of modern commercial lines; the wilderness has been interpenetrated by lines of civilization, growing ever more numerous. It is like the steady growth of a complex nervous system for the originally simple inert continent. (Turner 1961, 31–32)

The dominant idioms of this passage are of increasing complexity. While clearly an evolutionary tale of civilization and progress triumphing over primitive savagery, there is an imagined intimacy, naturalness, and mutuality in the imagery of *inter*penetration between American colonialists and aboriginal peoples. The romantic union of nature, wildlife, native peoples, and colonizers, suggesting their seamless complementarity, obscures the brutality through which the "slender paths" of indigenous peoples were "broadened" by "civilization in America." The "steady growth" of complexity out of simplicity and inertia appears to follow a pattern already etched into the soil.

In the same way that Anne McClintock argues that "imperialism cannot be fully understood without a theory of gender power" (1995, 6), but that "gender is not synonymous with women" (1995, 7), it could be argued that the idea of the frontier is *always also implicated genealogically*, for which *genealogical* describes the power to define and shape origins, or to organize generativity and vitality. From Turner's sexualized account of the frontier as an intimate zone of contact, the nation's body emerges as at once dominant, masculine, and all-encompassing. The mixing of organic imagery with that of "the complex mazes of modern commercial lines" portrays the nation's growth as both natural and inevitable.[23]

This frontier effect of technologically assisted generativity, in which progress is literally recorded as an advancing line and conceived as a process of civilization "filling in" the "arteries made by geology," could be described as a kind of magic semiotic compression, in which completely incoherent contents are congealed into recognizable form, and chaotic agencies become retrospectively original to the familiar. Repeated endlessly, the catechism of frontier lines disguises military occupation as "settlement," itself a term from geology, and produces the mythic transformation from a space of eliminated (autochthonous) nativity to the quasi-sacralized primal scene of the birth of a nation.[24]

Australia's First Frontier

Among the many contrasts that can be drawn between the colonial frontiers of Australia and the United States, one of the most prominent, but surprisingly little commented upon, is the role of sheep.[25] As the Australian historian Fred Alexander wrote in his account of Australia's "moving frontiers," "The earliest clear example of frontier influence is, of course, the pastoral frontier across the Blue Mountains made possible by the explorations of Blaxland and others and by the sheep breeding experiments of John Macarthur" (1969, 26). "Riding on the sheep's back," Australia's founding mythologies are distinctively ovine, combining the idioms of human and animal bloodlines in the forging of a new national identity. Even the frontier "line" of expanding settlement in New South Wales in the mid-1820s more closely resembles the scatter pattern of sheep on a hill than the advancing tides of Turner's imagined "filling in" of North America.

As the Australian geographer and historian T. M. Perry writes in his account of the spread of settlement in New South Wales from 1788 to 1829, much of the impetus to acquire new land came from the need for more grazing area, or pasturage. The successful crossing of the Blue Mountains in 1813 by Gregory Blaxland, William Lawson, and William Charles Wentworth is explained by Perry as a forced undertaking driven solely by the need to pasture livestock. He writes that

> the incentive that led to the successful crossing and, later, to the occupation of areas outside Cumberland was neither scientific curiosity nor need for land for new settlers. It was a loss in the stock-carrying capacity of the Cumberland Plain pastures that forced graziers to seek new areas to pasture their stock. . . . There was at this time no shortage of land in Cumberland upon which new settlers could be placed; in fact between 1812 and 1821 more than 228,000 acres of land on the Cumberland Plain were granted to settlers. The need was not for land but for grass. (1963, 27)

By 1825, Perry's map of the spread of settlement beyond the Cumberland Plains illustrates a settlement pattern composed almost entirely of sheep runs on lands held by "ticket of occupation" (a license to settle) extending west beyond Mudgee, Bathurst, and the Wellington Valley, marking what was then the limit of the initial frontier.[26]

According to historian Charles Burfitt's figures, more than ten thousand sheep had crossed over the Blue Mountains by 1820, establishing the inland sheep runs of New South Wales that almost immediately became the fledging colony's first viable export industry. In the same year, the quantity of wool shipped from Australia to Britain rose from 99,415 pounds to 175,433 pounds, a figure that would double by 1825.

In sum, it could be said that if the Australian economy was riding on the backs of sheep, the Australian frontier was trodden in by them. The "ever richer tide" of settlement described by Turner for America on the maps in which "settlement is colored," would in Australia have been the dirty white shade of Merino wool. Sheep, who graze more thoroughly than cattle, consumed the indigenous fauna in its entirety, including tubers, shoots, and seeds, as well as grass, terminating much of its subsistence capacity and converting it into mutton and wool. As Stephen Roberts describes this process, "From Bass Strait to the Pandora Pass, cattle and sheep were moving onwards; everywhere the lowing and bleating invaders were showing themselves a more relentless force of occupation than regiments of redcoated soldiers, and were passing over plain and mountain alike" (1964, 1). Like the Cheviot, whose breeding led to its praise by agriculturalists as "almost man-made," the selectively bred white tide of ovine settlement in Australia was itself a genealogical frontier, a product of the breeder's arts and an animal that literally embodied the modern industrial principles Australian settlement was intended to serve, uphold, and propagate.

A Tide of White Sheep

In the white tide of sheep that proved such a "relentless force of occupation" in Australia were united the bloodlines of improved sheep with those of their improving owners, as well as the flows of overseas finance and colonial ambition manifest as "animal capital" (Abbott 1971, 95). In turn, the "founders" or "fathers" of the Australian finewool industry, and in particular the figure of John Macarthur, have acquired an overdetermined paternal significance—not only as flock masters and squatter patriots but as nation builders, wealth generators, and heroic entrepreneurs. Even in their flaws and misdeeds, of which Macarthur was seen to have had no shortage, such men continue to be imagined as embodying the birth of Australian character and fortitude.[27] Thus another genealogical dimension of frontier mythology becomes apparent in its capacity to authorize, valorize, and genealogize paternity—rescuing the lost fathers through whose genius and inheritance the nation is seen to have grown and prospered.[28]

In the introduction to his 1913 essay titled the *History of the Founding of the Wool Industry of Australia*, opposite a signed picture of himself, Charles Trimby Burfitt, the honorary secretary of the Australian Historical Society, suggests that "the subject of sheep-raising is one that possesses more than a passing interest for the people of Australia." And yet, Burfitt chastises, although "the squatters, bankers, merchants, wool brokers, and others interested in the wool industry are, as a whole, a large hearted and generous people, [they are] *rather forgetful concerning events that have happened in the far away past*." Indeed, "few, if any, of those who have amassed and are amassing fortunes from the 'golden fleece' consider *who it was laid the foundation for their wealth*" (1913, 1; emphasis added).

In declaring his aim of setting "before the minds of the people of Australia" the magnitude of deficit accumulated through their collective nescience of national paternity, Burfitt is at turns loquacious and indignant. His lectures have as their primary object to remind his dilatory readers of their neglected "duty" to acknowledge "the debt of gratitude they owe to the memory of Captain John Macarthur for establishing . . . the fine wool industry, which has, more than any other, assisted to bring about the prosperity and important position Australia now enjoys in the world . . . and to remind them of their duty to the memory of the man who laid the foundation of [that] prosperity" (1913, 1). On behalf of his prodigal countrymen, Burfitt casts himself as a heroic narrator awakening their duty of memory toward a neglected patriot, patriarch, and pastoral pioneer.

This duty of national remembrance, like Turner's frontier origin story, is also a scene of national forgetting, as the social historian Bain Attwood suggests in his contention that white Australian historicism not only "informed racial opinion about the past, present and future of the country's aboriginal people from the beginning" but also that "European ideas and ideals about the course of historical change exerted a powerful influence upon conceptions of Aboriginality and in turn determined the rights of those defined as Aborigines" (2003, xi). As the postcolonial theorist Fiona Probyn has also argued (2003), the making of white Australian fatherhood and the unmaking of Aboriginal conceptions of land, lineage, and substance were combined in the birth of Australian nationhood, much as they had been combined in its occupation by white settler colonizers. As Attwood argues in *Rights for Aborigines*, the many native peoples who originally inhabited the continent "have only named themselves 'Aborigines,' 'blacks,' 'Kooris,' or 'Murris' etc. in the context of colonization" (2003, x). Aboriginality, dialectically constituted out of the need to become Aboriginal, and the need to be made so, thus emerges as an identity subordinated to the dominant genealogical forms of Europeanization, while in turn also being constituted by resistance to this condition (Trouillot 2003).

Macarthurism

Burfitt's view of the importance of John Macarthur is widely reproduced in the literature on the history of sheep and wool in Australia, and it can be considered one of the founding myths of Australian nation building, commemorated by Macarthur's portrait on the two-dollar note.[29]

Macarthur himself is the subject of a scholarly literature as voluminous as it is acrimonious—a contentious literary legacy that by most accounts, and even according to his most devoted hagiographers, comprises a fitting reprise to his confrontational personality. He is conventionally genealogized as a low-ranking British officer who arrived on the First Fleet with his wife Elizabeth and their infant son Edward, invested in sheep, built a farm in Parramatta, improved the flock by introducing Merino blood, and subsequently—through a combination of avarice, skill, persistence, luck, political influence, and, above all, a genius for self-promotion, used these improved sheep to attract English capital investment, thus contributing substantially to the founding of the Australian fine-wool industry.

On the subject of Macarthur's sheep there exists the same virulent disagreement that so often accompanies exalted patrilineage. In *The Merino: Past, Present, and Probable*, written in 1943, the agricultural

John Macarthur's portrait on the Australian two dollar note, next to a Merino ram, commemorates his celebrated role as a breeder-entrepreneur in the formation of the Australian fine wool industry, and his corresponding importance as one of the founding fathers of Australian nationhood.

historian Harry Braim Austin praises the Macarthur sheep, claiming that they "stood alone as the 'stud flock' of Australia" in 1820 and were in "great demand" (1943, 64). To underscore the reliability of his claim, Austin includes as an appendix an account from Ryrie Graham's 1870 *A Treatise on the Australian Merino* ("the most authoritative record we possess of early sheep-breeding in Australia"; Austin 1943, 218), in which both Macarthur and his sheep are highly praised:

> These sheep and their patriotic breeder constitute a chapter in the history of Australia [that is] important and interesting indeed to the whole world. From a worthless class of sheep Mr. Macarthur, with consummate skill and judgment, succeeded in producing sheep of the greatest value to commerce. . . . When Mr. Macarthur succeeded in obtaining the English Merinos from the Royal flock of George III there was not, perhaps, another gentleman of social position so distinguished in Australia that could have accomplished the task. (Graham qtd. in Austin 1943, 218)

Graham goes on to describe in detail the "Camden sheep," praising them as "undoubtedly the best in the colony": "It was in the year 1826 that I first saw these celebrated sheep; they were much smaller than the Australian Merino of the present day, and each had a small leathern collar around its neck. . . . The ewes would not have weighed more than 30 to 34 pounds each. Although small, they were extremely compact, low set, short-legged, bodies close to the ground, and as much alike as the grains in a handful of buckshot. Their wool was exceedingly fine" (Graham qtd. in Austin 1943, 218). While full of admiration, Graham's account is also qualified against the measure of improvement over time in Australian Merino sheep, particularly in terms of their size and the density of their wool. Hence, although Macarthur is praised for his "consummate skill and judgment," and the Camden sheep are described as having wool that is "exceedingly fine," they are also seen as in no way comparable to those of "the most distinguished breeders of Victoria at the present time," whose flocks "cast the best sheep ever bred by Mr. Macarthur into the shade" (Graham qtd. in Austin 1943, 218).

Graham's account also provides a glimpse into the glorified expectations of Macarthur's sheep in his acknowledgment that "like many a young Australian at the time, to believe in these sheep was part of my creed, and I certainly entertained no expectation of seeing them equalled [*sic*], much less excelled. . . . I know many squatters and persons connected to the squatting interests, when the Camden sheep were at the zenith of their popularity, who now assert that we never had or shall ever have, sheep equal to them" (Graham qtd. in Austin 1943, 218). In referring to Macarthur's sheep as "celebrated" to a "zenith of . . . popularity," and describing his own belief in "these sheep" as part of a "creed" shared by "many a young Australian," Graham suggests a combination of sheep faith and sheep hype that resembles the sheep mania described earlier by Stephen Roberts,[30] in which the improvement of sheep stock by skilled breeders such as Macarthur laid down the basis for Australia's emergence as a major economic partner in world trade, as well as a powerful modern nation-state. Both the Camden sheep and their flock master John Macarthur emerge from such descriptions as protomythic figures within Australia's imagined birth as a pastoral nation to which a series of foundational couplings are posited as originary—livestock and their breeders, Britain and Australia, and improved sheep and the creation of viable commercial markets.

It is exactly these mythic properties of Macarthur and his Merinos, epitomized by their commemorative reproduction on Australian currency, that are equally virulently decried by the anti-Macarthur mob, such as the author and historian John Garran. Like Austin a sheep farmer himself, Garran and his coauthor Leslie White's 1985 publication, *Merinos, Myths, and Macarthurs: Australian Graziers and Their Sheep, 1788–1900*, denounces Macarthurism as ignorant and misguided nostalgia.

According to Garran and White, Macarthur was nothing but a talented self-promotionist who remained ignorant of the most basic principles of animal breeding, overestimated the quality and value of his own flock, and used the idea of his superior purebred Merino sheep to accumulate land and other privileges from the government. As they claim of Macarthur's famously triumphal return to Australia from England in 1805, aboard the aptly named ship *Argo* and having procured stud animals from the royal Merino flock at Kew, as well as instructions from the British government to be given five thousand acres of prime farmland on which to increase and improve them, "The decision of Lord Camden, Secretary of State for War and the Colonies, implemented in 1805, to grant 5000 acres of land at the Cowpastures to John Macarthur and to indulge him with thirty convicts was against the recommendation of the Privy Council, the advice of Joseph Banks, and the land policy of Governors King, Bligh and Macquarie" (Garran and White 1985, 4). In other words, Macarthur had prevailed in his effort to relieve the fledgling colonial government of such an enormous parcel of their scarce and precious pasturage through shrewd and calculated application of the speculator's arts of spin, rather than because he was at all highly regarded in London. Macarthur's famous *Statement of the Improvement and Progress of the Breed of Fine Woolled Sheep in New South Wales*, delivered to the Home government while he was awaiting court-martial for assaulting a fellow officer in London in 1803, skillfully wove together the interests of his own farm and flock with those of the colony's future fortunes, with no other evidence to hand but a hank of wool.[31] Demonstrating a deft and persuasive line in product promotion, he embellished his wool sample with such glowing qualities and fabulous potential that it was all but money in the bank within just over a page of vigorous merchandizing. "The Samples of Wool brought from New South Wales," he (or his amanuensis) began,

> having excited the particular attention of the Merchants and principal English Manufacturers, Capt. Mc Arthur considers it his duty respectfully to represent to His Majesty's Ministers, that he has found, from an experience of many Years, the Climate of New South Wales peculiarly

> adapted to the increase of fine woolled Sheep; and that, from the unlimited extent of luxuriant Pastures with which that country abounds, millions of those valuable animals may be raised in a few years, with but little other expence [*sic*] than the hire of a few shepherds. (Macarthur 1803)

Uniting in his first paragraph the commercial "excitement" created by his "Samples of Wool" with the prospect of "unlimited . . . luxuriant Pastures" on which to raise "millions of those valuable animals," it is clear that Macarthur must have suffered greatly beneath his burden of duty "respectfully to represent to His Majesty's Ministers" what he had learned "from an experience of many Years" about the attractive ratio between expense and increase that could be realized from his fine-wooled sheep.

The samples of wool having been thus introduced as magical tokens of the economic bonanza to be found in the "peculiarly adapted" climate of New South Wales, their worth is further verified by their transformation into specimens confirmed by expert inspection to be possessed of "every valuable Property": "The Specimens of Wool that Capt. Mc Arthur has with him, have been inspected by the best judges of wool in this Kingdom; and they are of opinion, that it possesses a Softness superior to many of the Wools of Spain; and that it certainly is equal, in every valuable Property, to the very best that is to be obtained from thence" (Macarthur 1803). The superiority of the wool and its properties thus confirmed, and wedded to the prospect of unlimited increase, it is the sheep and their propagation that are next led into Macarthur's carefully staged statement of improvement and progress. These animals who are "of the utmost importance to this Country" are described as having shown rapid increase in their number, size, and weight, while at the same time their wool is seen to "visibly improve in quality" by measure of softness, fineness, and value. By crossing his animals with Spanish rams, he describes improving them still further, until he "far exceeded his most sanguine expectations," and at greater speed, calculating "that they will, with proper care, double themselves every two years and a half; and that in twenty years they will be so increased as to produce as much fine Wool as is now imported from Spain, and other countries, at an annual expence [*sic*] of one million eight hundred thousand Pounds sterling" (Macarthur 1803). Walking his Majesty's ministers through a

Statement of the Improvement and Progress of the Breed of Fine Woolled Sheep in New South Wales.

THE Samples of Wool brought from NEW SOUTH WALES, having excited the particular attention of the Merchants and principal English Manufacturers, Capt. Mc Arthur considers it his duty respectfully to represent to His Majesty's Ministers, that he has found, from an experience of many Years, the Climate of New South Wales peculiarly adapted to the increase of fine woolled Sheep; and that, from the unlimited extent of luxuriant Pastures with which that country abounds, millions of those valuable animals may be raised in a few Years, with but little other expence than the hire of a few Shepherds.

The Specimens of Wool that Capt. Mc Arthur has with him, have been inspected by the best judges of wool in this Kingdom; and they are of opinion, that it possesses a Softness superior to many of the Wools of Spain; and that it certainly is equal, in every valuable Property, to the very best that is to be obtained from thence.

The Sheep producing this fine Wool are of the Spanish kind, sent originally from Holland to the Cape of Good Hope, and taken from thence to Port Jackson.

Captain Mc Arthur being persuaded that the Propagation of those animals would be of the utmost consequence to this Country, procured, in 1797, three Rams and five Ewes: and he has since had the satisfaction to see them rapidly increase, their Fleeces augment in weight, and the Wool very visibly improve in quality. When Capt. Mc Arthur left Port Jackson, in 1801, the heaviest Fleece that had then been shorn, weighed only Three Pounds and a Half: but he has received Reports of 1802, from which he learns, that the Fleeces of his Sheep were increased to Five Pounds each; and that the Wool is finer and softer than the wool of the preceding Year. The Fleece of one of the Sheep originally imported from the Cape of Good Hope, has been valued here at Four Shillings and Sixpence per Pound; and a Fleece of the same kind *bred* in *New South Wales*, is estimated at Six Shillings a Pound.

Being once in possession of this valuable Breed, and having ascertained that they improved in that Climate, he became anxious to extend them as much as possible; he therefore crossed all the mixed bred Ewes of which his Flocks were composed, with Spanish Rams. The Lambs produced from this *Cross* were much improved; but when *they* were again crossed, the change far exceeded his most sanguine expectations. In *four* crosses, he is of opinion, no distinction will be perceptible between the pure and the mixed Breed. As a proof of the extraordinary and rapid improvement of his Flocks, Capt. Mc Arthur has exhibited the Fleece of a coarse woolled Ewe, that has been valued at Ninepence a Pound; and the Fleece of *her Lamb*, begotten by a Spanish Ram, which is allowed to be worth Three Shillings a Pound.

Capt. Mc Arthur has now about Four Thousand Sheep, amongst which there are no Rams but of the Spanish Breed. He calculates that they will, with proper care, double themselves every two years and a half; and that in twenty years they will be so increased as to produce as much fine Wool as is now imported from Spain, and other countries, at at an annual expence of one million eight hundred thousand Pounds sterling. To make

In his 1803 Statement to Parliament, John Macarthur emphasized his ability to improve fine-wooled Merino sheep through a series of successive crosses to become better adapted to their climate, and thus more productive. *Original document printed by Vacher and Son, Law Stationers, Parliament Street, London.*

few additional passages of agricultural detail and financial calculation "as a further confirmation of the principle of Increase," Macarthur arrives at his pitch. He agrees to bear all of the risk and expense of returning to New South Wales to devote his "entire attention" to sheep breeding for Britain in order "to accelerate its complete attainment." Requiring "no pecuniary aid," he asks only for "a sufficient tract of unoccupied Land to feed his Flocks" and the indulgence of selecting a few convicts to serve him as shepherds.

Despite the fact that his plan was opposed by Governor King of New South Wales, his successor, as well as Joseph Banks and several members of the Privy Council, Macarthur was able to secure support from Lord Camden and to acquire sheep from the royal flock. He returned to Sydney with these in 1805.

Elizabeth Farm

While opinion remains divided as to whether Macarthur's sheep were particularly significant to the emergence of the fine-wool industry in Australia, and indeed whether he was, as his most recent biographer, the historian Michael Duffy claims (2004), a man of honor, he continues to do figurative service for Australian colonial culture, character, and nationhood. His farm at Parramatta, now a museum, "reflects two centuries of European tradition and includes some of the oldest exotic plants in Australia" (4). Filled with copies of furniture, toys and portraits known to have belonged to John and Elizabeth Macarthur and their family, Elizabeth Farm is described in the Historic Houses Trust brochure for visitors as "a 'hands-on' museum" that "resembles a theatre-set" in which "props, stories and other theatrical devices are used to evoke and illustrate the family life of the Macarthurs." At Elizabeth Farm, half an hour by train west of Sydney and adjacent to its inland Central Business District (CBD), visitors are encouraged to wander about the house and gardens in order to revivify a sense of origins and discover their own sense of connection to the past by reaching out to touch, as well as see, the artifacts around them, in the real and reconstituted Macarthur family home.

In the same way visitors are encouraged to "wander freely" from one room to the next, there are innumerable connections to be made at Elizabeth Farm. While the Macarthur family tree offers an orderly and conventional version of the family's genealogical ties to present-day descendants, a number of other kinds of connection and discon-

The author at Elizabeth Farm. Elizabeth Farm at Parramatta, now a museum, is set on a scenic hillside with a 360 degree panoramic view of the surrounding countryside. Elizabeth Macarthur, for whom the property is named, managed the estate, its household, and its livestock during her husband's lengthy periods abroad. *Photograph by Sara Ahmed.*

nection drift in and out of the open doorways through which the family members once passed on a daily basis.[32] These historic corridors, now rebuilt and furnished with copies of the original objects, are filled with inquisitive tourists who are offered a curatorial genealogy of colonial history hybridized out of real, virtual, and synthetic components. So, too, as the short introductory video about the history of Elizabeth Farm points out, was the Macarthur family's own colonial history made up of partial, imagined, and lost connections. John Macarthur was absent from Australia for most of the crucial period between 1801 and 1817 during which his sheep were famously improved, and the operations of the farm in Parramatta, as well as other properties, were supervised by Elizabeth Macarthur, who also managed the family's flocks of as many as four thousand sheep.[33]

As Jill Ker Conway points out in her study of the Macarthurs and other early colonial New South Wales family sheep businesses, their success relied in no small part on the bifurcation of their family lives between home and colony. As she notes of the equally "pastorpreneurial" Riley family, their success

> was as much dependent on those of its members resident in England as it was on the work of the actual pastoralists in the colony. The achievement of success was reached with speed and efficiency precisely because it was a joint effort, carried out by a family whose members were resident on occasions both in Great Britain and Australia. This meant that they were able to bring a wide range of knowledge of their pastoral enterprises, and to adapt this knowledge in terms of the Australian environment with which they were all familiar. (1960, 220)

The dispersion of commercial responsibilities among family members positioned on either side of the British-Australian wool trade enabled genealogy to provide an entrepreneurial advantage, compensating through the closeness of family bonds for the enormity of geographical separation and the uncertainty of their financial prospects.[34]

These and many other accounts of the Macarthurs, their progeny, their home, their gardens, and their sheep reveal the connections and the fractures, the possessions and the dispossessions, and the visible and hidden histories of genealogy's subaltern channels, which, like the thick ivy tendrils covering its walls, cling to the Macarthur's Parramatta *cottage orné* as its verdant, antipodean supplement.[35] The making of the Macarthur family wealth—through land they acquired for sheep, whose selectively bred superior fine wool–producing qualities enabled John Macarthur to "spin" his sheep in London, where his son Edward provided him with crucial marketing strategies—illustrates the complexity of a transnational family genealogy bridging two ends of the wool trade and blending together sheep and human bloodlines in the service of Anglo-Australian enterprise. These blood ties enabled the generation of significant commercial wealth, lasting political influence within the colony, and the acquisition of vast amounts of land, establishing the Macarthurs as the leading patriotic figures of Australian settler society.

The attribution of the economic prosperity generated by the Australian fine-wool industry to the agricultural prowess and economic foresight of "original" colonial families, and in particular of patriotic, visionary forefathers such as John Macarthur, repeats and affirms a familiar ritual of genealogical reckoning, binding family narratives into the genealogies of nation-states, and retrospectively establishing a paternal discourse of nation creation in the language of father-

Visitors to Elizabeth Farm today are encouraged to explore the Macarthurs' domestic spaces in order to recreate a sense of quotidian life in colonial New South Wales. *Courtesy of the Historic Houses Trust of New South Wales.*

ELIZABETH FARM

Elizabeth Farm, 70 Alice Street, Rosehill 2142
Telephone 02 9635 9488 Fax 02 9891 3740
Web site: www.hht.nsw.gov.au

OPEN *Daily 10am – 5pm*
CLOSED *Good Friday and Christmas Day*

WELCOME

Elizabeth Farm is a 'hands-on' house museum. Visitors can wander freely through the house and garden, draw a chair up to a table, read family letters and newspapers, relax on shady verandahs or warm up before an open fire. With just enough furniture in each room to convey its function and character, the house is fully accessible: a museum without barriers.

The presentation of Elizabeth Farm resembles a theatre-set, where props, stories and other theatrical devises are used to evoke and illustrate the family life of the Macarthurs. As a place of learning, Elizabeth Farm aims to intrigue, raise questions, provide information and stimulate the imagination. Who were the Macarthurs? How did they manage their estate? What was it like to live here? Where did they fit into colonial society and why is this place important today?

Elizabeth Farm is a property of the
HISTORIC HOUSES TRUST OF NEW SOUTH WALES

The extensive reconstruction of the Macarthurs' daily lives at Elizabeth farm, where visitors are encouraged to reach out and touch a piece of living history, includes replicas of their children's toys, such as this four-wheeled sheep. *Photograph by Sarah Franklin.*

Joseph Lycett's iconic painting of Elizabeth farm, like many depictions of colonial New South Wales, offers an opulent panorama of civilization amidst natural abundance. Viewed from across the river by a well-dressed couple, the house is surrounded by a vast acreage and enclosed by a distant fence, emphasizing both the profitability and domesticity of successful cultivation. Hand-colored aquatint plate published 1 April 1825 by J. Souter, London. *Reproduced courtesy of the National Library of Australia.*

hood.[36] This generic genealogical work of nation generation noticeably relies on a depiction of shared substance through which the constitutions of sheep and men are combined to found nationhood in enterprise.[37] The work of genealogy being performed through John Macarthur and his sheep is also that of strengthening the lineages that link Australia to Britain through shared blood and wool, shared ancestry, and shared commerce. These powerful symbolic connections are condensed and performed through overlapping idioms of paternity—from Macarthur's authorship of a proper Australian breed of sheep to his patriotism in doing so to the establishment of his patronym of one of Australia's founding families.[38] Even the disputes about his importance confirm his enduring role as a quintessentially Australian father figure.

As with all genealogical narration and affirmation of this kind, the foregrounding of some lineages requires the suppression of others. The missing genealogical background to the John Macarthur story includes both the importance of his wife's contribution to their family's agricultural and entrepreneurial achievements (although, somewhat unusually, the farm is named for her), and the violent history of Australian settlement through which the original inhabitants of Parramatta were replaced by white families and their sheep. As in John Lycett's lyrical portrait of colonial New South Wales as a pastoral idyll, in which a couple and their home together occupy a vast and tranquil natural landscape, the genealogy of John Macarthur is naturalized as an origin, or new beginning, figured as a father.

While noting that an equally vast quantity of genealogical dropout is required to achieve this effect of a new beginning, as in Lycett's painting, where the couple stand, somewhat like an innocent, newly arrived, colonial Adam and Eve, it is important also to notice the magic of genealogical thinking that allows this effect: for it to be both singular as a beginning, and continuous over time as multiple lines of descent.

As we have seen both in the work of Frederick Jackson Turner, where the idiom of the frontier as a crucible is used to describe how mixtures of settlers, soil, toil, and contact with indigeneity become original and heritable identities ("Americanization"), so in Australian frontier mythology squatter culture and the outback ethos are depicted as the germ cells or protoplasm *out of which a new genealogy is born*.

New Breeds

One of the most striking legacies of the distinctive idioms discussed in this chapter—of genealogy, the frontier, and paternity—is their reaffirmation in the context of Australian reproductive biomedicine at the culmination of its second century.[39] These developments not only confirm the shaping influence of genealogies of the past upon the future but the underlying principle of this connection, or transfer, which is the means by which genealogy establishes a direction. The genealogical orientation of the transfer of sheep science into IVF was not only familial in the sense of creating new families but also in its combining of sheep and humans on a new frontier. Like earlier successful breeding initiatives, Australia's first IVF births were depicted as patriotic feats of scientific achievement and as a source of national pride. As Anthony Fisher writes in his celebratory account of Australian IVF, "bicentennial Australia leads the world in many aspects of the reproductive technology 'race,' having produced many IVF 'firsts,' much of the early embryo experimentation, the first legislative regulation, and so forth" (1989, 6). And referring to the birth of Australia's first test-tube baby, Candice Reid, the world's third IVF child, born on 22 June 1980, he writes, "the success of the Australian teams was widely publicized and became a matter of national pride: of the first 16 test-tube babies, 3 were born in Britain, 1 in the U.S., and 12 in Melbourne" (1989, 12). As well as a measure of scientific excellence, and world-leading technology, these achievements, as Fisher suggests, stand as evidence of enhanced national fertility, an accomplishment Fisher expresses in language evoking the nation as progenitor: "Australia prides itself," he claims, "*on having produced* over two thousand 'test-tube babies' and several world firsts: frozen embryo babies, donor egg babies, IVF twins, triplets, and quadruplets" (1989, 4; emphasis added).

The story of these achievements is narrated in considerable detail by the philosopher and science studies scholar Harry Kannegiesser in his bicentennial publication *Conception in the Test-Tube: The IVF Story; How Australia Leads the World* (1988). Once again, this story is very sheepishly Australian. As Carl Wood notes in the opening sentence of his preface to Kannegiesser's book, the story of IVF is once again that of man and sheep united on an unknown frontier: "The excitement of seeing a sheep embryo under the microscope triggered the development of in vitro fertilization in Melbourne" (1988, x).

Wool to Embryos

This excitement belonged to Alan Trounson, one of Australia's most eminent scientists and the director of the country's newly established Stem Cell Bank at Monash. At the age of sixteen, Trounson accepted a scholarship from the Wool Board of New South Wales to become one of the wool technology students pursuing wool technology subjects at the New South Wales Institute of Technology in 1963. He graduated in 1968 and went on to pursue a master's degree at the Agricultural Research Station in Hay in western New South Wales, where he completed his dissertation on the promotion of multiple births through selective crossbreeding, entitled "Reproductive Characteristics of Merino and Border-Leicester Ewes" (1968).[40]

Based on the strength of his experiments, and fueled by an increasing interest in embryology, Trounson was accepted by Neil Moore

Pride in pastoral achievement continues to be celebrated at Sydney's annual Royal Easter show, now held on the grounds of its 2000 Olympic Games, which offers the opportunity to admire large numbers of "champion sheep," whose genealogies, like those of the families who bred them (and whose names they often share) also comprise a celebrated feature of Australia's British heritage. *Photograph by Sarah Franklin.*

of the University of Sydney Department of Animal Husbandry, an expert in large animal surgery and "Australia's leading scientist in agricultural reproductive biology" (Kannegiesser 1988, 342). Moore, who was trained at the Reproductive Biology Unit at Cambridge and who is said to have been the first to suggest the idea of IVF to Carl Wood in 1969, supervised Trounson's PhD on the development of fertilized sheep ova, completed in 1973. Following his doctorate, Trounson secured a postdoctoral fellowship to join the Agricultural Research Council Unit of Reproductive Physiology and Biochemistry at Cambridge, where he undertook a series of studies of control of oocyte maturation, embryo freezing, and embryo transfer, and worked with numerous prominent and soon to be prominent scientists including both Neil Moore's supervisor, Robert Moor, and his fellow postgraduate Steen Willadsen, the first scientist successfully to clone sheep using somatic cell nuclear transfer.

Having spent four years at Cambridge studying cryopreservation of livestock embryos, the ova and embryos of cattle, horses, rabbits, rats, and pigs, the hormonal cycles of recipient and host ewes, and the chemical and molecular events of early pregnancy, Trounson was perhaps uniquely equipped as a scientist to return to Melbourne in 1977, at the request of Carl Wood, to assist him in the development of IVF.

From his extensive training with livestock embryos, ovulation induction, and ovine uteri, Trounson was able rapidly to introduce improvements to IVF, although not in time to win the race against a neighboring Melbourne team, who delivered Australia's first IVF success in 1980. Nonetheless, Trounson's contributions to IVF were soon to gain worldwide recognition and acclaim, including his introduction of hormonal stimulation to induce superovulation, which has become standard practice in IVF. Trounson also pioneered techniques of embryo culture, embryo transfer, embryo biopsy, and embryo freezing, as well as sperm microinjection, oocyte and embryo donation, and oocyte maturation.

Continuing in a long line of Australian founding fathers born of sheep and wool, Trounson stands as Australia's most celebrated reproductive pioneer, working on the frontiers of biomedicine to improve methods of assisted reproduction and, more recently, to direct Australia's efforts in stem cell propagation. His achievements thus extend the frontier ethic of improvement literally back in to genealogy, affirming the ongoing centrality of agricultural reproductive biology to the life sciences, as well as their commercialization. The importance of the sheep-human interface, historically so formative in Australian history, remains clearly evident in the ongoing technology transfer that intertwines sheep and human reproduction. The curious colonial return of stem cells made in Australia to London described in chapter 2 retraces the path of a voyage that has become increasingly, not less, well-described as a lifeline.

Commonwealth Biofutures

As I noted at the outset, the aim of this chapter has been to examine the importance of genealogy and the frontier as constitutive technologies of colonial and national emergence and identity as these comprise both essential histories of ovine "biological control" and an indication of the extent to which these were as much global as they were local, economic as much as biological. In asking what questions Dolly poses for the future remaking of genealogy, now itself a scien-

Fine wool continues to be one of Australia's signature products in terms of both quality and quantity. The science of improving sheep has been generously funded by the Wool Board, established in 1937 and formerly the Woolmark company, now Australian Wool Services Limited, one of the world's most influential wool promoters. *Photograph by Sarah Franklin.*

tific frontier, it is useful to consider the genealogical orientations that contextualize the horizon, or dawn, of her creation. Thus while sheep pose an unconventional standpoint from which to consider colonialization and its legacies, their ongoing significance to scientific innovation, reproductive biomedicine, and emerging biomarkets such as stem cells make such a vantage point almost unavoidable.

Like Roslin, Dolly is a direct descendant of the Imperial Bureau of Animal Breeding, established in London between the wars in 1929 to consolidate the vital agricultural connections built up over more than a century between Britain and her colonies and former colonies. Similarly, Dolly's birth was prefigured in more than a century of exchange of sheep, sheep products, sheep science, and sheep improvements between Britain and her colonies and former colonies, in particular Australia.[41] Dolly continues these connections in ways the chapter has sought to explore not so much in terms of how her individual genealogy can be traced back through particular lines and ties of descent, but rather how in principle we might begin to think more pointedly about how the remaking of genealogy is orientated in particular directions, with consequences for whose worlds, lives, and hopes will build the future.

There are thus several ways in which it proves useful to think of Dolly as an animal that belongs to Britain's imperial heritage of colonial expansion, while also being a key figure in the promotion of Britain's leading role on the contemporary frontiers of bioscience. For example, one of the most important questions posed by Dolly's lineage is that of the relationship between biological expansion, transnational biological traffic, and control of genealogy, or, more particularly, what this chapter calls *economic genealogies*. As this chapter suggests, Dolly is both conceptually and constitutionally the viable offspring of more than two centuries of continuous trade between Britain and Australia based on sheep experimentation, sheep breeding, sheep products, and exchanges of actual sheep. In this manner the genealogical mixtures out of which she is made commingle lineages of science and industry with those of agriculture and medicine in ways that take their shape not only from regional and national but also imperial and colonial and indeed global histories. Hence the importance of idioms of genealogy to the promotion of national interests and the production of national identities are explored in both this chapter and the next to argue that it is all but impossible to disaggregate the actual and the metaphorical dimensions of Dolly's lineage as a national, imperial, or colonial animal. To the extent that reproductive biomedicine and stem cell science are frequently envisaged in terms of future promises, it is useful to consider their historical connections to much earlier patterns of capital accumulation, selective breeding, and transnational partnerships in the creation of new products, markets, and trade. Much as Joseph Banks's and Robert Bakewell's purebred experimental flocks could be seen as scientific animals in sheep's clothing, so, too, were such rationalized and re-engineered animal bodies the conduits for the international exchange of specialist knowledge about breeding and improvement, consolidated through both commercial and scholarly ties. It is the intersection of these complex historical ties with particular lines of industry captured in the genealogies of animals such as Dolly that this chapter has analyzed as a set of vital economic, scientific, national, colonial, and biological relations.

The most challenging example of how genealogies are created

and remade, and for which this chapter provides only a descriptive scaffold, is the way in which the Anglo-Australian legacy of shared sheep and wool bonds was transferred into human IVF in the 1980s, when the standard protocols for IVF were adapted to include ovulation induction, a technique originally developed for sheep (Robinson 1967).[42] We could consider Trounson, who cloned a lamb as a doctoral student at the University of Sydney by splitting a six-day old morula in two, as much a product of a distinctive Anglo-Australian scientific coupling based on sheep and wool as Dolly, Roslin, Clomid, or stem cells. The fact that Trounson was supported by an Australian Wool Board Scholarship to study in Cambridge,[43] where his contemporaries included many of the scientists who later contributed to Dolly's creation (and together comprise one of the world's most powerful and accomplished scholarly lineages of reproductive biologists), cannot be considered purely coincidental, but rather forms part of a legacy this chapter tries to begin to unpack. United by historical ties of commerce, industry, and science, British and Australian scientists can be seen to share a lineage of vital innovation within the life sciences, bonding them, in Haraway's terms, as kindred offspring of a specific mode of "sociotechnical production" (or, in this case, *re*production; 1997, 7). The family tree that links the sheep experiments of Wilmut and Campbell to those of Trounson and Wood, and back further still to those of Bakewell and Banks is historically rooted in a shared commitment to agricultural improvement, livestock industrialization, medical and scientific progress, and the generation of new commercial markets.[44] This profitable intercourse is now manifest in projects such as tissue engineering, human (and ovine) genomics, gamete exchange, and reproductive biomedicine—which allow us to see that Dolly's coming into being was not only a specific technical accomplishment but formed part of a much larger pattern of sheep experimentation that continues to have world-historical dimensions.

Like the voyages of Joseph Banks, whose efforts to nurture his favorite colony were inseparable from his concerns with the improvement of sheep,[45] the sea crossings of British and Australian scientists left in their wake a vast collection of findings, both material and theoretical, through which natural history might be put to work in the service of empire (Jardine, Secord, and Spary 1996; Miller and Reill 1996). The legacies of these contributions to what Patricia Fara has called "imperial science" (2003, 152), or Richard Drayton names even

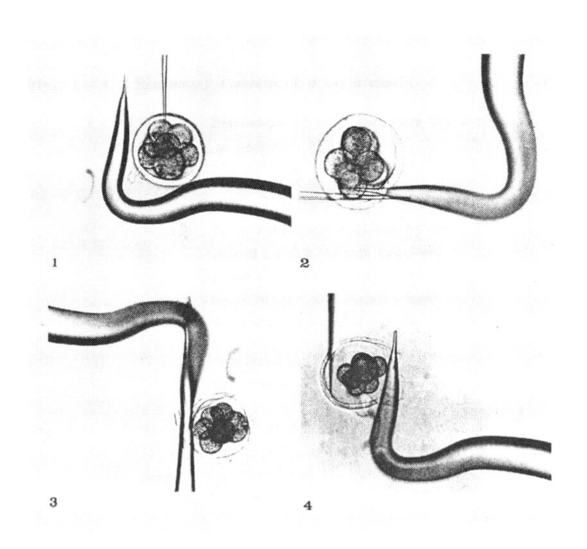

Plate three of Alan Trounson's doctoral dissertation at the University of Sydney illustrates the use of special "crook" pipettes to puncture and remove a portion of the zona pelucida, or protective shell, of an eight cell sheep blastocyst. *Reproduced by permission of Alan Trounson.*

more precisely "the agrarian view of empire" (2000, 103), are not only to be found in Kew Gardens (where Banks kept his famous flock of Merino sheep) but in the carefully tended gardens of wealthy north Sydney suburbanites, where English tea roses bloom and Jack Russells bask in the tropical heat. So, too, are they evident in more recent exchanges, such as the vials of stem cells "born" in Trounson's Monash laboratories and sent traveling back to Britain, via Singapore, in a kind of postcolonial biological handshake with the scientists at King's.

A consequence of such a view is to resituate the cloning question and the Dolly technique in the history of mixtures of agricultural, commercial, medical, scientific, and industrial purposes with which experimental sheep breeding has been linked for centuries. This context allows the thickness of Dolly's genealogy to become a site of critical reflection on the many connections and continuities she embodies, and likewise emphasizes the disjunctures, boundaries, and separations that, in a sense, cloning puts into crisis. As in the history of nation and soil Dolly embodies as a Scottish sheep, her links to genealogies of imperialism and colonialism cannot be separated from their legacies of violence and death, often mobilized under the banner of improvement and a better life. I explore the question of death that inevitably accompanies any form of selective breeding further in the following chapter, which also examines the idea of "biological control."

This chapter has approached the work, or thickness, of genealogy not only as a form of linkage, shared descent, or shared substance but as a motivated economic consolidation of interests that can itself be given direction, orientation, and even purpose. Genealogy in this sense is not merely a matter of socially constructed nature or a bit of pruning on the family tree. The ways that genealogy can be imagined to originate on a frontier demonstrate its powerful imaginary dimensions—but these are hardly "just" in the mind. Seen in this light, the sheepishness of Australian nation formation, like its frontiers, can be revealed to have several different faces, or directions, that widen the cloning question to become more world historical, while also reconfiguring familiar histories of nation, empire, and colony by exploring their pastoral dimensions. In turn it is possible to examine the highly motivated, interested, constitutive, organized, strategic, and cumulative ways in which genealogy is directed toward certain ends, and how this process of genealogical orientation is both reproductive and selective.

5 Death

> I've got a motley crew of animals. They will not get them. I've got chains, superglue, vicious geese, barricades and I'll lock myself up with them in my kitchen.
>
> —Cumbrian farmer during the British foot-and-mouth epidemic of 2001

> The origins of the Herdwick breed are shrouded in mystery. A well-known story tells how the breed was established in the Lake District from a flock of 40 sheep washed ashore from a Norwegian vessel wrecked off the Cumberland coast in the tenth century. Herdwick sheep are now only found on Lake District farms.
>
> —Susan Denyer, *Herdwick Sheep Farming*

> I will make an arch in Scotland from Locharbriggs red sandstone. It will be erected in a sheepfold on the Lowther Hills near to where I live, after which it will follow a drove route from Scotland through Cumbria and into Lancashire or Yorkshire. This is one group of work within the sheepfolds project which will have its origin and destination outside Cumbria yet still leave its mark there, in common with the people, animals and things which have passed though this area over the centuries, leaving evidence of their journey but neither coming from nor staying there.
>
> —Andy Goldsworthy in Goldsworthy and David Craig, *Arch*

Dolly the sheep was euthanized on 14 February 2003 "because of the presence of a virally-induced lung tumour resulting in progressive decline in respiratory function," according to "Dolly: A Final Report," her official autopsy published in 2004 (Rhind et al. 2004). She was five years and seven months old, and had given birth to six offspring. The arthritis she developed in 2001—in her left patella, distal femur, and proximal tibia—was unusual, but the ovine pulmonary adenocarcinoma (OPA) of which she was dying when she was euthanized

is caused by Jaagsiekte sheep retrovirus (JSRV), an infectious disease commonly transmitted among sheep housed closely together in barns through airborne water droplets. Thus, although Dolly died, in effect, of a common sheep disease, and according to the authors of her autopsy report is "no reason to think Dolly was more vulnerable to infection because she was a clone" (2004, 156), it is also the case that her arthritis proved exceptional and that her telomeres were found to be shorter than those in four control sheep, thus confirming the earlier observation that her telomeres differed at one year from those of age-matched controls (although Wilmut and his colleagues caution that these measurements were made in several different experiments and may be misleading). They also note that although Dolly was in many respects a perfectly normal sheep, and that "during routine husbandry there were no unusual findings apart from the development of her arthritis," the cell culture from which her nucleus was transferred had been through twenty-seven population doublings and that "since the birth of Dolly, experiments in several species have revealed a range of abnormal phenotypes associated with unusual patterns of gene expression" (2004, 156).

Thus, although Dolly's ability to reproduce confirmed her health, normality, and vitality, her death was a reminder that she embodied new mixtures of mortality and immortality, normality and pathology that add to the many paradoxical dimensions of her significance. Above all, she instantiates the paradox of "biological control" described by Ian Wilmut in his warning that the age of biological control will be characterized by new forms of responsibility for scientists, governments, and regulators who must decide how far that control will be extended into the intimate architecture of life itself.

When I visited Roslin in July of 2000, one of the first conversations I had with Wilmut was about Dolly's eventual death. Beset by suspicions about her health from the outset, international scientific scrutiny of her biology was all but inevitable, but cast a kind of shadow across her future, as if everyone expected her to die at any moment. Wilmut did not seem to be so troubled by the possibility that Dolly's health might raise questions about the safety or viability of cloning by nuclear transfer, although he is one of the leading scientists to call for a scrupulous international register of all such findings in all cloned animals in order for the science of cloning to be evaluated as thoroughly as possible. On a more personal note, however, he clearly rued the day she might no longer continue her happy presence among her companions at the institute. "People think I wouldn't miss her because I could make another Dolly," he said, "but what they don't understand is that she has her own individuality and there would never be another sheep like her."

Inside the security-controlled corridors of Roslin are photographs of its charred remains in the wake of a firebombing in the 1980s that destroyed much of the facility. "We keep these photos here to remind us that what we do, and how we do it, really matters to people," Wilmut said. The shock he experienced when Roslin was so violently targeted by animal rights activists is one of the driving forces behind Wilmut's concern about "biological control," how it will be used, and how the public will participate in decisions about its future. British regulations governing experimental procedures on animals are the oldest and strictest in the world, dating from the Cruelty to Animals Act in 1876, and currently under the Animals (Scientific Procedures) Act 1986, which specifies, among other things, that no animal can undergo more than one experimental procedure. Nonetheless, and as the legislation makes explicit in its requirement to document the deaths of all animals resulting from experimental procedures, the ties that bind together animal and human suffering are the very same that link medical scientific experimentation to the hope and promise of improvements in human health.

For these and other reasons, Dolly's life as a clone and an experimental animal, though celebrated as a miraculous success, was shadowed from the beginning by her death to such an extent that there was surprisingly little comment when she actually died. Indeed, many people thought she was already dead. Throughout the research for and writing of *Dolly Mixtures*, I was surprised how often people expressed their condolences to me: "I was so sorry to hear that Dolly died," they said, never able to recall exactly how or where they had learned of her demise, when I informed them that she was not only alive but now had several offspring. As we saw in the earlier discussion of Damien Hirst's lost sheep as both sacred and tragic, the broad pastoral idiom of care for the flock is also paradoxically composed of the protection of animal life and control over animal death.

This chapter explores the tensions within pastoralism and domestication both as a further means of contextualizing Dolly's life and death and as a means of exploring further the idea of 'biological control." Using two highly politicized contexts of lost sheep, from Australia and Britain, as case studies, it is possible to map some of the wider contours of the deathly foreshadowing that haunted Dolly's life from a slightly different angle. What was notable about the public grief expressed in 2003 about the Australian sheep trapped aboard the cargo ship *Cormo*, unable to disembark at their intended Saudi port due to allegations of disease, was both its impassioned affect and its complex connections to nationalism, agriculture, and global politics. The question of why sheep death can generate such complicated forms of national mourning and sentiment was even more dramatically demonstrated in Britain during the foot and mouth outbreak of 2001. Here again, as in the *Cormo* episode, a contagion lay at the heart of the crisis, generating contagious connections that led in all sorts of directions, demonstrating how global sheep continue to be and how their mobilities or movements continue to shape national economies. Threatened by foot and mouth were precisely the lifeways and rural traditions seen to epitomize the British countryside, and hence the politics of the countryside that emerged in the wake of the epidemic affected everything from the tourist industry to the food chain.

As this chapter demonstrates, the highly emotional reaction to the contagious sheep episodes described here is in some ways surprising, especially given that sheep are cheap, low-status livestock animals with terminal futures. Exploring the "affective economies" (Ahmed 2004) surrounding contagious sheep is thus a useful way to track the complexity and force of the underlying connections linking people and sheep through shared genealogies of blood and soil, as well as local and national identities. Also striking is the manner in which these episodes lay bare the fundamentally economic basis of sheep life and sheep death—a reality contagious diseases make explicit, as they are inevitably more harmful to balance sheets than to animals or humans themselves (foot and mouth is no more severe than the common cold to a sheep and is rarely fatal). The viability of sheep is thus inextricable from their economic utility, but as we shall see, the question of where sheep economies begin and end can be as difficult to trace as the bramble of affect surrounding them. In the same way that Dolly's death was fortuitously almost coincident with the liquidation of her parent company, PPL Therapeutics, so too it is not irrelevant that the first colony of human embryonic stem cells was successfully derived in London at around the same time, in the early Spring of 2003. The linkages between these events are not direct, but they point to a dynamic this chapter tries to emphasize, namely the ways in which the making of new life is difficult to disentangle from the production of new forms of morbidity (a principle the Dolly technique illustrates particularly well), and the importance of the negative form of biological control, which is its loss. In asking to what extent sheep contagion is a matter of "biological control," when the logic of such control is essentially economic, not biological, we return to a dynamic that is fundamental to Dolly's viability as an offspring, and inheres in the very term "livestock," which is that of the intertwined investments of agriculture, science, medicine, "improvement," and economic growth. As it turns out, the intense emotions surrounding these investments are a revealing species of affect as well.

The Ship of Death

On August 5, 2003, the Dutch-owned carrier ship MV *Cormo Express* left Fremantle, Western Australia, bound for Saudi Arabia. On board was her live cargo of fifty-eight thousand sheep, destined to supply the lucrative Middle Eastern halal market in the month of Ramadan with lambs, hoggets, and wethers "turned off" from wool production into Australia's buoyant live-export industry. Worth more than 200 million Australian dollars annually, the expanding export market in live animals plays a key role in the growth of Australia's highly competitive meat and livestock industry.[1]

The export market in live animals, mainly sheep, from Australia has come to play an increasingly vital role for its rural industries as they seek new means of responding to the vicissitudes of the global economy by shipping as many as 6 million live sheep overseas per annum. Much of the expansion in the sheep and lamb meat export trade from Australia since the late 1990s has occurred through diversification to include a wide variety of markets through which it can sell its lowest-value cuts. According to figures compiled by the United States Department of Agriculture Foreign Agriculture Service (USDAFAS) for 1998, Australia exported 10,159 tons of lamb meat to Papua New Guinea, largely in the form of "low-trade brisket and flap product." In addition to "low-trade" meat, exports of uneconomical animals, such as Merino wethers (two-year-old males) from Western Australian wool flocks, have found a niche market in the foreign live-animal trade.[2] Because the Merinos are grown for their wool, their mutton is considered poor quality and largely unacceptable, thus making it suitable only for the very "bottom end" of the cheap-meat market for meat consumed, for example, by the large Saudi population of domestic servants who are almost without exception unskilled immigrants living on subsistence wages.[3] Wealthy Saudis consuming "high-end" meat, in contrast, have created yet another market—for the much sought-after mutton of fat-tailed sheep, now also raised in Western Australia for export alongside their less gourmet shipmates ("sheepmates").

Australia remains the world's largest producer of sheep and wool for export. According to the USDAFAS, Australia had 116.9 million head of sheep in 2000—a ratio of approximately ten sheep for every human Australian, and more than twice as many sheep as New Zealand, its closest competitor. Just under half of Australian sheep are lambs (45 million), of which a third (16 million) are slaughtered annually. In 1999, Australia's fresh and chilled lamb meat exports rose by 47 percent in response to declining U.S. and South African sheep meat production, among other factors. This rapid growth has in part resulted from the Australian government's ongoing support and assistance and the establishment of initiatives such as the Lamb Industry Development Programme,[4] aimed to facilitate the sheep industry's gradually increasing emphasis on meat production.

Live sheep exports from Australia are almost exclusively shipped to the Middle East and North Africa, with Kuwait, the United Arab Emirates (UAE), Jordan, and Saudi Arabia acting as the main importers in a trade relationship active since the 1950s.[5] While Saudi Arabia is supplied by sheep and lamb from several African countries including Somalia, Ethiopia, Kenya, Tanzania, Zimbabwe, Djibouti, and Yemen, its largest supplier of sheep and lamb meat is Australia, followed closely by New Zealand. Australia exported 6,811,565 sheep in 2001, the country's peak year to date in a trade that has doubled in volume since 1990.[6]

According to the Australian Meat and Livestock Agency,

> A major reason for the strong demand for live sheep in these countries is the traditional preference for freshly slaughtered sheep meat, handled in accordance with Islamic religious beliefs. The growing preference for younger sheep and the Middle-Eastern fat-tailed sheep have been catered for by Australian producers. An increasing number of young

> sheep (lambs and hoggets with two permanent teeth) are being turned-off for live export and the introduction of fat-tailed sheep varieties into livestock production, particularly in Western Australia, has lifted exports. (Australian Meat and Livestock Agency, 2004)

Demand for live sheep from Saudi Arabia is fueled by its high population growth (largely through immigration) of as much as 3.75 percent per annum (during the 1990s). No duty is levied by the Saudi government on imported livestock, to the benefit of both trading partners. Particularly during the Muslim holy month of Ramadan, when an estimated 2 million pilgrims travel to Mecca to perform Hajj rites, for which live sheep and lambs are in high demand for ritual feasts and gifts, the volume of imports both expands and reaches as much as a 40 percent higher price.

Refugee Sheep

When it set sail in August of 2003, the livestock vessel *Cormo Express* was resuming a trade with the Saudis that had not infrequently been stymied by conflict in the past. Between 1990 and 1996, Australian exports to Saudi Arabia ceased entirely as a result of disputes between trading partners. Hence, on August 21, when the *Cormo*'s cargo was rejected by Saudi inspectors in Jeddah, who claimed that 30 percent of the flock were suffering from the viral infection scabby mouth, a well-worn historical battle was reinstigated over the question of sheep shipment quality—a dispute often enveloped in a miasma of contagion allegations and attempts to prove their falsity.

The officially acceptable level of sheep mortality on an overseas journey such as that made by the *Cormo Express* is currently considered in Australia to be 2 percent. More than a 2 percent mortality rate provides grounds for an investigation and the possible suspension of export licenses. In New Zealand, from where increasing numbers of sheep were exported to Saudi Arabia during the 1990s as a result of Australia's withdrawal from live-export trade relations with one of her largest intermittent partners, stringent measures of quality assurance are adhered to as a means of ensuring the acceptability of shipments.

The *Cormo Express*, then, whose departure from Australia was a sign of economic health, shortly thereafter departed from Jeddah amid an epidemic of controversy and under an unwelcome denomination as the "ship of death." These events were quickly followed by the collapse of the live-export trade between Australia and Saudi Arabia, effected by a ban on further shipments. The refusal by Saudi officials in Jeddah to accept the shipment was followed by a storm of rejections from other countries—even of offers from the Australian government to deliver the shipment for free. Following intervention by the Red Cross, a concerted diplomatic effort was launched to find a destination for the lost sheep, more of whom were perishing in the heat on a daily basis. Thus while animal rights protestors urged a slaughter at sea of the suffering livestock, John Howard, the Australian prime minister, revealed in an emotional broadcast that more than twenty countries had so far been approached to accept them to no avail, and that political efforts to disembark them had been redoubled.

Repatriation of the flock afloat was widely rejected as an option by representatives of Australia's 11 billion-dollar meat industry, who feared contamination of domestic flocks and who joined a raging debate that took on increasingly mainstream political significance. As David Hawker, a Victorian rural MP for the Liberal Party, was quoted in the *Sydney Morning Herald* on 15 October 2003, "Bringing the sheep back to Australia after potential exposure to exotic diseases or insects not found in Australia could turn a very difficult situation into a national disaster" (qtd. in Metherell 2003, 6). Despite being tested by an internationally accredited veterinary team who found no evidence of scabby mouth disease among the sheep on board the *Corma*, and in spite of protestations that the animals had been vaccinated twice against the disease, making its appearance highly improbable, industry representatives and MPs protested their return to Australia. The Australian Meat Industry Council, cautioning that market access and international confidence in Australian meat had been built on its extreme diligence maintaining strict quarantine procedures, released a statement in which it claimed that "the return of these sheep after more than two months at sea would risk shattering the confidence in Australia's largest food export industry" (qtd. in Metherell 2003).

In time, with diplomatic assistance and after more than fifty countries had refused to accept the *Cormo* sheep, Eritrea was paid 1 million Australian dollars to receive them, along with a donation of three thousand tons of feed.[7] More than six thousand sheep had died after more than ten weeks at sea—a morbidity exceeding 10 percent of the original shipment, and thus five times higher than the maximum acceptable levels, leading to a ban on further shipments to Saudi Arabia and thus the demise of this formerly lucrative trade.

While uncertainty continued to surround the Saudis' rejection of the shipment, the health of the sheep on board, and the fate of the *Cormo*'s ovine cargo, John Howard reaffirmed his government's support of the live-export industry, arguing that it was an important trade that contributed to the livelihoods of many rural Australian farmers. Yet in addition to disease, such export shipments remain vulnerable to the vicissitudes of global trade, such as the sharp rise in the Australian dollar coinciding precisely with the *Cormo*'s voyage and presumed by many Western Australian farmers to have been the actual reason for the sheep's rejection in Jeddah.[8]

Border Control

The transformation of a live-export shipment from a signifier of economic health into a scandal of animal death was not only played out on the world stage as a diplomatic crisis and tragedy of animal suffering but also as a complicated reprise on Australian politics of nation and repatriation in the context of heated debate over the country's immigration policy. In 2001, John Howard's highly controversial decision to refuse entry to the Norwegian ship *Tampa*, carrying 460 refugees from Iraq and Afghanistan rescued at sea from their sinking vessel, was widely seen to have increased his popularity, and strengthened his reputation as a protectionist leader. In the crisis of the imperilled Australian sheep at sea two years later, the Howard government's position was again protectionist, inevitably raising awkward comparisons. As in the *Corma* case, public sentiment toward the *Tampa* refugees lost at sea had focused on the need to protect Australia's economic health by refusing to accept would-be immigrants whose desperation was seen by some only to confirm the danger of allowing others to follow in their wake. Allegations that refugee parents had attempted to throw their own children overboard were refuted by close observers of the rescue effort, who claimed that such distorted imagery was intentionally circulated to dehumanize the refugees and to enhance the depiction of them as threatening foreign bodies. Two years later, amid vivid media coverage of the desperate *Corma* sheep attempting to throw themselves off the ship to escape their stifling cesspool, compassion for the animals rose to such heights that protestors in London donned sheep suits to protest outside Australia House on the Strand.

Back at home, while Australian pride in the country's prominent meat industry suffered in the face of accusations of diseased animals and the international rejection of its livestock, the very quarantine laws seen as the guarantor of Australian meat quality required that Australia reject its own animals. The crucial dispute over the question of whether the Australian sheep were diseased remained unsettled and unsettling, while the shipment was repeatedly rejected by one country after the next in an abject scene of international haggling. Forced to disown their sheep for fear of contaminating the meat industry, large sectors of Australian public opinion and a tearful prime minister remained determinedly pastoral in their care and concern for the innocent and vulnerable animals—very much in contrast to the views of those same constituencies two years earlier, when they decried the struggling *Tampa* refugees as dangerous, threatening, and undeserving.

In the notorious *Tampa* affair, Howard defended his decision to reject the asylum seekers rescued by the Norwegian carrier ship, and to detain them at sea until an alternative destination could be hurriedly brokered or bought, in the name of maintaining Australian security by forcibly protecting the country's borders at the risk of attracting negative international publicity.[9] In his view, a line needed to be drawn around Australia, keeping foreign refugees and foreign contagions alike at bay. The government represented quarantine and immigration control as protection against unwanted, and potentially dangerous, intruders. This was exactly the tragedy the *Corma* cargo reversed when its sheep fell victim to the very same quarantine logic, leaving them stranded on the outside of the line Australia drew around itself in the name of protecting its own.

In 2001, Howard graphically depicted the asylum seekers aboard the *Tampa* as illegitimate and manipulative, implicitly implicating them as contagious vectors to terrorism (Lygo 2004).[10] In what many saw as a deliberate electioneering tactic, Howard's hard line on asylum seekers was aimed at appealing to a popular xenophobia in which Australia had become a "soft touch" for the world's poorest migrants and vulnerable to infiltration by terrorist sympathizers.[11]

Howard's emotional patriotism in relation to the stranded Australian sheep aboard the *Cormo Express* thus precisely mirrored, but caricatured, the intensity of his antiasylum invective against the desperate passengers aboard the *Tampa*. In both cases, the core of his argument about security was primarily economic, and its mechanism was spatial: both the asylum seekers and the lost sheep threatened Australia's economy by breaching the borders it needed to protect itself against outsiders. But the contrast between national mourning over the deaths of humanized Australian sheep and national apathy toward the lives of de-humanized non-Australian people raised disquieting contrasts. In these instances of what Judith Butler (2004) describes as "precarious life," the value systems that make of some lives innocent, heroic, and deserving subjects of public mourning and grief and of others the threatening, ungrievable others, are starkly drawn against a familiar backdrop of sacrificial sheep.

Contagious Connections

The contagious connections emanating from the sheep aboard the "ship of death" thus once again unite issues of global trade, rural industries, economic growth, and patriotism. What Al-Jazeera news dubbed the "politically embarrassing Australian sheep," and the BBC referred to as the "sheep of shame," became a new kind of Australian refugee—at once mourned within, but excluded from—the homeland to protect its borders against exotic contamination. The Australian trade minister Mark Vaile denounced the sheep ship episode as a "fiasco," while the Australian press called the "sensitive" and highly secretive deal with Eritrea a "Red Solution," denoting both the Red Sea port of Massawa where the sheep were disembarked and the so-called Pacific Solution to the *Tampa* affair.

At the center of the *Cormo* controversy was scabby mouth disease, or contagious ecthyma caused by the parapox virus. Although rarely fatal to sheep, and unlikely to infect humans, the disease is notable for its rapid speed of contagion. To assess the effects of the viral vectors, or contagious connections, established through sheep diseases is to encounter the role of the biological in classically agricultural form and to witness also some of the most ancient forms of "biological control," such as the isolation of diseased animals from a herd. However, what is also surprising about these biological connections, especially in the context of an increasingly complicated and global politics of the food chain, is their scope, demonstrated by their capacity to link together unexpected elements into new patterns that reveal a different order of things. Hence, Australia's "gift" to Eritrea, like a kind of Trojan Sheep, comprised an act of martial economics, in which sheer economic power enabled the Australian government to protect its commercial interests by staging a humanitarian, face-saving end to a multimillion-dollar trade fiasco at sea. *Contagion* in this context acquires added significance, intertwining as it does the danger of an unseen infectious enemy with the danger of economic exposure.[12] The contagion itself was never confirmed, and it may never have existed, as the ship's veterinarian claimed. By contrast, the contagiousness of the episode, and its ability to spread rapidly, more accurately describes the political and mediagenic character of the controversy than its reputed source. From the start an economic venture, the "ship of death" primarily fell victim to a failed commercial transaction, the remedy to which was to desist in commercial relations of this kind altogether.

Sheep-Watching

For contagion to be at the heart of the dispute between Australia and its Middle Eastern trade partner, over a cargo of sheep in the prelude to Ramadan, and for the dispute to exist in the elusive realm of shifting and unverifiable claims and counterclaims, surrounded by rumor, propaganda, and spin, is generically consistent with many conflicts in the global era in which modern, rationalist accounts fail to achieve purchase on their objects and become, instead, parodied by their own inadequacy. Under these conditions, the heightened sense of technological potency associated with modern industrial design, scientific advancement, precision engineering, and so-called smart weapons is seemingly mocked by forces so far beyond its control, or even comprehension. It is in such cases that even the most technologically "advanced" nations often must resort to almost premodern equipment: in this case sandbags, trenches, sniffer dogs, and fire pits. This primitivism, argued by Karl Marx to be the primary default mode of capitalism, may increasingly in the contemporary era be seen not as something modernity ever leaves behind, but as its enabling and continual precondition. In an era in which even so-called natural disasters, such as droughts and floods, are described as man-made, the difference between progress and it opposite can be elusive.

This is one of many reasons sheep serve as such useful bellwethers for a present-day global watch. An ancient animal central to the very idea of capital, economy, and wealth, and as instrumental to the foundation of ancient pastoralism as to modern industrialization, sheep travel in between world systems as readily as they cross the world's oceans, pass imperceptibly into goats, or undergo IVF. Sheep thus remain significantly global animals whose economic value continues to be generated through international exchange and whose commercial protection is considered a matter of national security.

Both the ongoing centrality of sheep to the modern British economy—a prominence wrongly presumed to have become a quaint historicism—and the importance of their paths, or movements, were powerfully demonstrated during 2001, in an epidemic of foot and mouth disease (FMD) that bridged the transformative events of 9/11 with a crisis of animal health that starkly contrasted the idea of "biological control" against its opposite in the form of uncontrollable contagion. In the confused response to FMD—by one of the world's most scientifically advanced, wealthy, and powerful nations—is evident a combination of extreme measures and intense emotionality that calls into question the very meaning of biological control by providing a case study of its opposite—an out-of-control biological disaster. In the interstices of the ensuing debate about biosecurity were the connecting circuits of a host of other security concerns—from worries about discarded airline food to the British army's diet of Argentinean beef.

From the point of view of modern animal husbandry, the danger of contagion is that no one can really know if it is there or not, and so protection measures must not only be rigorous but preventative and, above all, certain.[13] Protection against contagion, by definition an inexact science since by the time it is noticeable it has already gone out of control, must be highly efficient, definite, reliable, and complete. It is for this reason that contagion begets eradication, the only way to be completely sure. To the industrial agriculturalist, contagion does not allow for half measures: it is a natural either/or, that is, either spreading or eliminated. As the tenth edition of the Australian livestock handbook *Sheep Management and Their Diseases* advises in its one-sentence entry under "treatment" for one of the most feared and highly contagious viruses in sheep, foot and mouth disease: "There is no satisfactory treatment for the disease *nor would any attempt at the same be permitted* should the disease make its appearance in Australia or New Zealand" (Belschner 1976, 490; emphasis added). Foot and mouth is not only a disease that divides the world's nations, like the world's sheep, neatly into the haves and the have-nots but a contagion that is notoriously virulent and difficult to control. By virtue of their geographic isolation and stringent quarantine requirements, Australia and New Zealand are among the only FMD-free countries in the world. Protection of this status is considered a national priority of such importance that, as H. G. Belschner, the author of *Sheep Management and Their Diseases* and former chief of the Division of Animal Industry in the Department of Agriculture in New South Wales advises, "the control of foot-and-mouth disease is a matter for Government Authorities, both Federal and State. Because of its dramatically infectious nature, early recognition of the disease and prompt reporting of suspected cases are of vital importance in its control and eradication" (1976, 490–91).

FMD UK

The foot and mouth crisis that began at Heddon on the Wall in Northumberland in February 2001 quickly spread to sheep flocks throughout Britain, opening up new corridors and pathways for infection across its length and breadth. Six months later it continued to spread into new areas, threatening to breach the top of the Lune valley, from where it could sweep back south down the Lancashire-Yorkshire border from Cumbria—one of the worst affected areas in Britain and the long-acknowledged backbone of its complex sheep-breeding system.

By September, official figures claimed that 3.8 million animals had been slaughtered, with 10 to 20,000 awaiting culling and disposal on a daily basis. On top of heavy losses for the farming industry, the so-called plague of foot and mouth had severely damaged the tourism industry and cost the taxpayer an estimated 10 billion pounds in lost revenue, compensation, and expenses related to control of the disease. It was widely feared that many parts of the countryside would never recover the lifeways and livelihoods linked to sheep farming that had existed for centuries in Britain and which have given the British countryside many of its prized distinctive features. Chief among these were the distinctive qualities of rural life—not only the picture-postcard pastoralism of areas such as the Lake District but the farm and village economies that had sustained families over generations, which might no longer exist once the disease was eliminated. If it could ever be eliminated, that is, for, as the authorities also widely acknowledged, it might as easily become endemic, not only in Britain but throughout Europe, as was the case before 1967, the year of the last major outbreak before the thirty-four-year intervening period created what many realized too late had become a tragically over-complacent sense of security.[14]

From the beginning, the foot and mouth crisis divided the country profoundly. It was difficult for people outside rural or semirural parts of the country to comprehend what a pall the epidemic cast over huge but sparsely populated areas of Britain, particularly in the north. From cities such as Lancaster, in northwest England just below Cumbria, and less than an hour south of Longtown—the epicentre of the disease and its nominal ground zero—it was impossible to avoid a sense of crisis and the eerie discomfort of an invisible but constant contagious threat. The lethal but mysterious spread of the disease through odd quirks, such as its attraction to rubber, created a disturbingly vague yet visceral sensation of uncertainty, fear, and dread. The lugubrious sense of foreboding was intensified by the clamp-down on human, animal, and vehicular movements that might contribute to the spread of the disease. It became difficult to know how to act responsibly under such bizarre and threatening conditions, when even riding a bike along familiar back country roads suddenly became an act of shocking carelessness. Visiting friends in the countryside became both an important pilgrimage of support and a worrying cause of possible infection, creating conflicting impulses that led to complicated conversations about everything from the future of rare-breed Lakeland sheep to the date of the next national election. Civic duty, at once urgently called for and at the same time ill defined and ambiguous, existed in a particular kind of tension with the sense that almost anything could happen next in the unnaturally still and quiet countryside.

People openly made comparisons to the wartime conditions that required constant vigilance about the minor details of daily life, like saving milk caps. Moving from this kind of quotidian angst down to a meeting in London or a conference in Brighton exposed the deep divisions between those for whom the future of the rural was an everyday cause of unavoidable concern and those for whom the foot and mouth saga was an incomprehensible overreaction to the deaths of farm animals destined to be slaughtered anyway.

What was difficult for anyone without experience of a rural econ-

DAILY EXPRESS

THE BIGGEST WEEKEND PACKAGE IN BRITAIN

SATURDAY MARCH 24, 2001 50p

120 GREAT PAGES PLUS FREE MAGAZINE INCLUDING 7-DAY TV LISTINGS

ZETA LINED UP FOR NEW ZORRO FILM

EXCLUSIVE: SEE PAGE 15

BECKHAM'S DATE WITH DESTINY

SPORT BACK PAGE

GREAT EXPRESS OFFER: 60% OFF FLY DRIVE HOLIDAYS SEE PAGE 71

CARNAGE

25 million animals are facing slaughter as foot-and-mouth epidemic turns into a national catastrophe

FULL DRAMATIC STORY AND PICTURES: PAGES 6&7

LETTERS 58 GARDENING 62-64 KIDS' PAGE 65 CROSSWORD 68 TRAVEL 75-85 CITY 98-102 SPORT 103-120

Foot and mouth disease (FMD), although highly contagious, is not a serious illness for sheep, and rarely affects humans. The slaughter of millions of sheep in Britain in 2001 resulted from the economic division between FMD-free and FMD-affected countries—a distinction that is primarily economic. *Reproduced by permission of the* Daily Express.

omy to understand was the apparent paradox that it is because animal lives are valued that their deaths must be properly managed. Foot and mouth upset the careful balance between animal life and death at the heart of a rural ethics that combined constant proximity to animals and devotion to their care (husbandry) with management of their transformation into useful products to be sold at a profit, something that by definition means ending animal lives prematurely. Control of the foot and mouth epidemic, which required the destruction of millions of animals—including ancient lines of stock owned by generations of families and irreplaceable herds of carefully bred varieties of sheep and cattle—upset the balance of the rural economy in almost every aspect. Few people would have expected the twenty-first century to dawn with such a forceful demonstration of the ongoing importance of animal bloodlines to the lifeblood of the nation's economy. Neither were the complex emotions concerning animal life and death a familiar scene on the front pages of the nation's newspapers, catalyzing an impassioned debate about our intimate connections to livestock.

A particularly painful irony of the unprecedented slaughter of millions of British cattle and sheep was that foot and mouth has never posed a risk to human or animal health. Unlike the new form CJD, or so-called mad cow disease, foot and mouth is harmless to humans and rarely infects them. It does not even affect sheep very much and is frequently depicted as an ovine version of the common cold. Foot and mouth is only lethal to domestic animals because it is *economically intolerable* to humans. Because the world market for sheep, pig, and cattle products is divided between countries where foot and mouth is endemic and those officially designated as disease free, the dilemma of foot and mouth has always been an economic one. Still a further irony is the high cost of remaining disease free, and therefore more marketable—an expensive means of trade protection, as Britain quickly discovered.

Since 1967, Britain had remained successfully disease free through what became known as the slaughter policy—a policy it shared with its former island colonies of Australia and New Zealand—whereby any infected herds were immediately culled. For geographically isolated nations, the slaughter policy was thought to offer a competitive advantage, although this view had always been seen to rely on a rather precarious footing. The view that Britain's rural sheep economies were safe from foot and mouth in part because they were remote from sources of infection constituted one of the main assumptions disputed throughout the attempt to control the disease.[15] The debate about vaccination, for example, centrally turned on this question. The worst-case scenario, which many in government and in the sheep and wool industries feared, was that even after aggressive and prolonged use of the slaughter policy during the 2001 outbreak, the government would be forced to turn to vaccination anyway because the disease would have become endemic.[16]

Foot and mouth, while not a lethal disease, then, is nonetheless a lethal form of infection. It is precisely the highly contagious nature of the disease that is seen to necessitate the strict market separation between diseased and disease-free sources of meat and other animal products. However, as in the case of the sheep aboard the *Cormo Express,* whose suspected infectious status was provided as a justification for their rejection, contagion, like security, is both a condition in itself *and* a means of producing other conditions—such as martial law, quarantine, or the slaughter policy. The management of foot and mouth in Britain provided a case in point of managing many other things, including the future of the rural, the politics of the countryside, the tourist industry, and the food chain. The map of the closed footpaths, back roads, and carriageways crisscrossing Britain's evacuated countryside reversed the image provided by Frederick Jackson Turner of human paths built on animal trails, which then became roads, establishing the network of commercial activity associated with civilization. Foot and mouth paralysed modern transport systems, returning the paths to the animals and closing down access to the countryside, to markets, and to farmyards, as well as bringing to a virtual halt Britain's vast commercial trade in sheep and sheep products.

Eventually the army was called in to manage the "chaos" of foot and mouth. The attempt to extract the disease from the countryside, through the combination of a slaughter-on-suspicion (SOS) policy and a three-kilometer mandatory cull zone around any infected animal, produced scenes of brutality and destruction that were frequently described as barbaric. At the same time, other scenes from the epidemic were as ominously modern, such as the teams of men in white coats from the Ministry of Agriculture, Fisheries, and Food (MAFF) mobilizing a highly coordinated nationwide exercise in animal testing, documenting, classifying, and processing. Foot and mouth was a mixture of extremes. It demonstrated how thoroughly the rural and the global are matted together into the fabric of sheep's wool, how profoundly the nation's economic lifelines remain rooted in the same soil that gave birth to the industrial revolution in the north of England, and how the very same intensities of circulation that still bind them together are also conduits for their mutual undoing.[17]

To examine these connections, which reveal the ongoing centrality of the ancient sheep economies to what might be called the new postmodern pastoralism of the countryside, through which sheep breeding has been linked to tourism, organic meat products, rural conservation, and the heritage industry, the following section analyzes a series of newspaper headlines from the height of the foot and mouth epidemic—in March, April, and May 2001—to ask what kinds of connections were being made along the infective pathways established through the disease and its spread. While the footpaths of rural Britain remained closed for over a year, new pathways creating unexpected connections were opened up by foot and mouth, often framed in terms of the management of life and death, and often expressed as intense emotion. These unexpected connections and "af-

Brown puts his foot in mouth

● Farmers' fury at plan to kill one million healthy animals ● Minister causes alarm with cattle gaffe

The sheep depicted in the *Daily Telegraph*'s coverage of the dreaded exponential leap in cases of FMD are a sturdy group of Cheviots whose solidarity and dignity are reinforced by the absence of any humans. *Reproduced courtesy of the* Daily Telegraph.

fective economies" once again demonstrate the thickness of the remaking of animal genealogies as forms of rural or national industry, or capitalization.[18]

The image shown above from 16 March 2001 comes from the *Daily Telegraph*, Britain's leading conservative broadsheet and the traditional newspaper of the conservative, or Tory, party. The morality of the *Telegraph* is emotional and traditional, often bordering on nostalgic, with strong strains of nationalism or even neo-imperialism often showing through the seams of its no-nonsense editorials. This staid and stalwart British newspaper responds to everything from stem cells to immigration issues with a visceral, almost instinctive, "shire Tory" conservatism as far removed from the monetarist neoconservatism of Margaret Thatcher as from the liberal New Britannia of Tony Blair. During the foot and mouth crisis, William Hague, a Yorkshireman, served as the leader of the Tory party.

The second week of March saw the dreaded exponential leap, from 22 to 256 suspected cases of foot and mouth infection, confirming the worst fears within the farming community of an uncontrollable epidemic. It was on the evening of 15 March that Tony Blair's agricultural minister, Nick Brown, announced the draconian scorched-earth policy of a mandatory three-kilometer culling zone around all confirmed cases of FMD. Under the front page headline of the next day's

Telegraph, "Brown puts his foot in mouth," the paper ran a photograph of a small group of Cheviot sheep, standing in a group on an open plain.[19] The image is both dignified and poignant. The sheep are looking in different directions, underscoring the headline's critique of the government's confused agricultural policy. The photograph is compelling precisely because it makes sheep, often regarded as timid and subordinate farm animals, look almost heroic. The animals are standing firm, as if in solidarity, while their fate is determined in faraway corridors in Whitehall. They look alert, capable, and fit—in "rude health"—as well as innocent and harmless. The lead sheep is shot in profile, gazing into the distance, while others meet the viewer's eye directly, face to face. It is a flattering image that gives dignity to the animals, together on their land, with no humans in sight. Their motionless poses appear statuesque beneath the open sky, at what appears to be a sunset on the open plain.

The image is also distinctive in its depiction of scale. It is unusually cropped to run the full breadth of the newspaper page, offering a panoramic expanse. This wide-screen presentation intensifies the effect of a wide-angle lens that at once magnifies the open land and sky overhead but also veers somewhat eerily out of perspective. On second glance, or even subliminally, it becomes apparent that there is something not quite normal about this otherwise very ordinary panorama of sheep grazing in a field, again underscoring in visual terms both the ominousness and the scope, as it were, of the looming catastrophe. The photograph is emotional in part because it is simple, and in part because it is more complicated than it might initially appear. Its editorial instincts on the pulse of the countryside, as ever, the *Telegraph* foregrounds the tragedy of ministerial ineptitude, the victims of this folly, and the "farmer's fury" that has ensued through the bodies of sheep—their stoicism implicitly invoking that of the endlessly-trodden-upon rural farmer. These are the core elements of the foot and mouth crisis, and they are deeply affective in ways that saturate the intensely commercial bodies of sheep with everything else those commercial lines knit together. What does not need to be said by the *Telegraph* is that the bloodlines of sheep and those of the rural country lifeways they make possible are as good as one and the same.

The ancient ethos of blood and soil that sheep continue to epitomize, and which make them the object of such intense emotional in-

● One million healthy animals to be killed ● Change in tactics could save May 3 poll date

Ministers gamble on mass cull

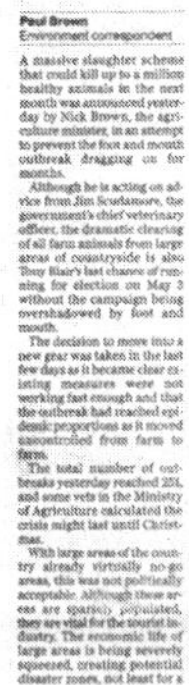

Paul Brown
Environment correspondent

A massive slaughter scheme that could kill up to a million healthy animals in the next month was announced yesterday by Nick Brown, the agriculture minister, in an attempt to prevent the foot and mouth outbreak dragging on for months.

Although he is acting on advice from Jim Scudamore, the government's chief veterinary officer, the dramatic clearing of all farm animals from large areas of countryside is also Tony Blair's last chance of running for election on May 3 without the campaign being overshadowed by foot and mouth.

The decision to move into a new gear was taken in the last few days as it became clear existing measures were not working fast enough and that the outbreak had reached epidemic proportions as it moved uncontrolled from farm to farm.

The total number of outbreaks yesterday reached 251, and some vets in the Ministry of Agriculture calculated the crisis might last until Christmas.

With large areas of the country already virtually no-go areas, this was not politically acceptable. Although these areas are sparsely populated, they are vital for the tourist industry. The economic life of large areas is being severely squeezed, creating potential disaster zones, not least for a

Triplet lambs born yesterday on a farm near Longtown in Cumbria. Though healthy, they are destined to be culled under the government's new action plan Photograph: Murdo MacLeod

Labour's lead up to 26 points

Anne Perkins

Labour's lead over the Conservatives has extended to 26 points, the biggest lead since 1998, according to a Gallup poll in today's Daily Telegraph. It puts Labour on 55%, up 6 points from last month, with the Tories unchanged on 29%. The Liberal Democrats have slipped to 12%.

But support for a May election appears to be ebbing. A poll for Channel 4 News last night showed only a third of voters backed May, down from two-thirds less than a week ago.

English local elections must formally get under way on March 26. To postpone them legislation would have to be rushed through parliament next week. The Home Office insists there are no plans for delay.

The Conservative leader, William Hague, commenting on the Channel 4 poll, said: "It is a sign of how seriously people take this crisis and I think it is quite right that they do so."

The sheep on the front page of the *Guardian* appear somewhat distorted, as is the message of the photograph, which verges between the comical and the bizarre. *Reproduced courtesy of the* Guardian *and Murdo Macleod.*

Government warns that hundreds of farms must be cleared of all livestock to contain foot-and-mouth

One million lambs to the slaughter

By Nigel Morris, Ben Russell and Steve Boggan

MORE THAN one million sheep and pigs will have to be slaughtered if the Government is to have any chance of containing the foot-and-mouth crisis, the National Farmers' Union said last night.

As farmers prepared for an unprecedented cull of livestock, mostly sheep, the Government warned that hundreds of farms in a huge swath of land from Cumbria to southern Scotland will be cleared of all livestock susceptible to the virus.

Tony Blair sanctioned the move after he met farmers' leaders in Downing Street on Tuesday when they told him he faced the "most difficult decision" he would ever have to take. Mass culls will eliminate any flock that has been in contact with the three markets at the centre of the outbreak. Animals linked to two of the largest livestock dealers in Britain will also be killed.

Nick Brown, the Minister of Agriculture, said at least 150,000 animals could be put down. The initial cull will exclude uninfected cattle because they are least likely to spread the disease, Maff said.

But, privately, farmers' leaders said the toll would be far higher than the government's estimate. A senior official of the NFU said: "The number of animals involved could be phenomenal. It is

Maurice Bowman feeding sheep at his farm in Kirkoswald, Cumbria: 'You can only push people so far. I don't want to kill my animals' Warren Smith

INSIDE

- Further reports PAGES 6&7
- Racing rescheduled PAGE 24
- Leading article REVIEW, PAGE 3
- Michael Brown REVIEW, PAGE 4

beleaguered tourist industry in those areas. Mr Brown said: "This is a policy of safety first. We are intensifying the slaughter of animals at risk in the areas of the country – thankfully still limited – where the disease has spread."

Eighteen new cases were confirmed yesterday – all of them in areas already infected – bringing the total number to 251. About 161,000 animals have been slaughtered, with 64,000 waiting.

Jim Scudamore, the Government's chief veterinary officer, said the disease had initially been transmitted by animal movements through markets, but drastic action was needed to stop it spreading from farm to farm, and moving down valleys, because the virus was carried on the wheels of vehicles and the boots of workers.

Many farmers are strongly opposed to the slaughter, with some farmers in almost open revolt. At Kirkoswald in Cumbria, Maurice Bowman, a sheep farmer, warned that the plans would force many to breaking point. "You can only push people so far," he said. "I don't want

The sheep on the front page of the *Independent* announcing the FMD cull are shown with a farmer and his vehicle, emphasizing the close ties between rural farmers and their flocks, and the tragedy of the mass slaughter policy for rural ways of life as well as livestock. *Reproduced courtesy of the* Independent.

vestment despite the fact they are simply "dumb beasts" destined for the slaughter house, is a legacy neither the left-wing *Guardian* nor the left-wing leader Tony Blair showed much aptitude in comprehending. The aptness and nuanced affect of the *Telegraph* image contrasted significantly with the coverage of the same story provided by both the *Guardian* (the Labour paper) and the *Independent* (the smallest, Liberal, national paper). Under "Ministers gamble on mass cull," the *Guardian* offered another wide-angle image of a group of sheep, but one that conveys a much less coherent set of messages and, consequently, a less well-defined point of view. Looking distinctly peculiar, a lamb-bearing ewe, probably a Swaledale mule, from Longtown is shown with her three offspring in front of a man, a dog, and a truck. The man is frowning ambiguously at the sheep and holding onto his dog, while the sheep look into the camera unexpressively. The caption reveals that "though they are healthy," these sheep are "destined to be culled under the government's new action plan"—suggesting it is the vulnerability of young lambs the picture is emphasizing.[20] But the message is confused as these lambs are not photographed to be cute, but oddly, with a fish-eye lens, so that they look distorted. Too many relationships occur in this image so that it becomes more of a composite of elements of rural life (man, sheep, dog, vehicle) than a pointed message about their fate, its injustice, its tragedy, or its origins in ministerial ineptitude.

Similarly, while we can attempt to make sense of the image on the front page of the *Independent* under the headline "One million lambs to the slaughter," it is, like the *Guardian* photograph, confused and lacking focus. The cloth-capped farmer stands surrounded by his animals, feeding his flock. Behind him are his dogs and his four-wheel-drive vehicle. The tragedy depicted is presumably that he will have to kill his sheep, but the image does not stamp itself with authority on the viewer, as neither the farmer nor the sheep evoke a sense of dignity. The farmer appears tired, and the sheep mill about his feet hungrily in a photograph that nonetheless conveys a tragic sense of unity between the farmer and his flock.

The combination of humans and animals in both the *Guardian* and the *Independent* photos is also what confuses their message, detracting from their emotional power and thus shifting more emphasis onto their strictly literal interpretation. What is revealed by comparison to be very effective about the *Telegraph*'s image is that the *absence of humans* enables sheep to take on a more human aspect themselves, and literally to stand in for the tragedy that will affect both humans and sheep. This theme of the rural domestic animal as a heroic figure representing human grief was widely evident in press coverage of the foot and mouth crisis, and it raises a number of key questions about the organization of pastoralism, its affective economies, and the meaning of animal domestication as an extension of human relationality.

The means by which sheep were used to represent the saga of rural life and death, its beauty and tragedy, its ebbs and flows, was primarily evident in the context of foot and mouth in the way sheep were used to express a sense of loss and victimization. This became repeatedly evident in the depiction of farmers' inconsolable grief at the mandatory—and to many, arbitrary—culling of their flocks. The arrival of FMD in early spring, just before the lambing period, added significantly to the sense of tragedy, both because of the transformation of the life-giving period of the sheep's yearly cycle into its opposite, and because the disease worsened steadily into the Easter period, so that by late April, the associations with the paschal lamb had become heavily overdetermined. As if anticipating this unfortunate sequence of events, the *Telegraph* coverage of foot and mouth on 17 March pictured a Cumbrian farming couple kneeling with their soon-to-be-slaughtered lamb between the headlines "We would rather die than let them kill our flock" and "Church bells will ring out to show their support." The tenant farmer Josephine Wheatley, who built up her flock of twelve hundred animals over twenty years with her husband Brian, is quoted as saying, "the compensation doesn't matter. They are our lives." Not having had a vacation in twenty years and threatened with homelessness if they lose their flock, Mr. Wheatley protests that "our sheep are part of our family and to stand by and let the government kill everything we've worked for is absolutely devastating."

Similarly illustrating the intensity of farmers' grief at the loss of their flocks, and the age-old injustice of the cosmopolitan government's callous disregard of the farming community, the case of Anne Young, also living near Penrith, raised a national outcry after she was pictured on the front page of the *Daily Telegraph* on 17 March with one of her "very rare wild merino sheep" named Susie.

Dressed appropriately in a light brown fleece, and wearing a depressed expression of grief, Anne is depicted in a Madonna-like as-

79p
Saturday
March 17 2001
Published in Manchester
and London
guardian.co.uk

The Guardian NORTH

The news from Ground Zero: foot and mouth is winning

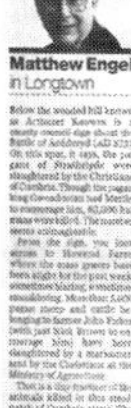

Matthew Engel
in Longtown

The Daily Telegraph

www.telegraph.co.uk
Britain's biggest-selling quality daily
Saturday, March 17, 2001 75p

Farmers in revolt over mass cull

Action group declares 'all-out war'

'These merinos are my family'

(top) The dramatic imagery in the media during the FMD crisis was puzzling to some observers, who could not understand why the culling of animals already destined for the slaughterhouse was cause for so much concern. For others, such images could not begin to capture the destructive effects of FMD throughout the rural community. Reproduced courtesy of the *Guardian*. *(bottom)* In emphasizing that her rare-breed Merino sheep are "her family," the Cumbrian farmer Anne Young of Penrith underscores the intertwined genealogies of farmers and their flocks, encompassing both the labour of establishing a top-quality line of animals, and a shared sense of belonging to a particular place or soil. *Reproduced courtesy of the* Daily Telegraph.

pect holding her threatened sheep. A vegetarian opposed to killing animals, she is quoted as saying, "I have a gun, but the police tell me that if I think of using it I would be arrested and the cull would continue without me. I intend to be with them if they have to die, since I have been at nearly every one of their births."

Another Cumbrian farmer is introduced in the *Independent* of 16 March, along with his gun: "Maurice Bowman has a gun but he swears he will not use it when the men from the Ministry call. He will just tell them to leave him in peace with his sheep and cattle. 'I'll be polite, and I'll leave my gun in my cupboard,' he said. 'But I'm not sure others will be so restrained. You can only push people so far, and I'm afraid we've cracked.' " Pictured next to one of the infamous burning pyres that received so much news coverage around the world over the course of the epidemic, and especially during March and April, Bowman is quoted citing an old Chinese proverb: "The man with the full stomach has many problems, but the man with an empty stomach only has one." In these and hundreds of similar stories, human and animal lives were bound together in ways that were depicted as both emotionally powerful and deeply tragic. The threat of the loaded weapon pointed not at sheep but by the farmer at himself, or a member of the ministry, is repeatedly invoked, indexing an intimate interdependence that can only be comprehended as a matter of life and death. The gun—a complicated emblem of extreme emotions, rage, masculinity, and death—is almost always depicted alongside elaborate and disturbing declarations of love for animals and grief at the loss of livelihood that has at its heart a very particular kind of human-animal bond.

The Sunday papers of 18 March identified Longtown in Cumbria as the epicentre of the outbreak and the ground zero of the epidemic.[21] It was in Longtown, in early February 2001, that an infected consignment of sheep from Hexham had been distributed all over Britain. Ten percent of all initially infected farms were in the Longtown area, which was within less than two months almost entirely devoid of sheep.

Formerly Britain's largest sheep market, Longtown would normally have seen twenty thousand head of fat stock change hands on a daily basis—mostly hoggets, or year-old lambs reared on the hills, and largely destined for meat suppliers in the south. Longtown, where almost everyone is involved in sheep farming in one way or an-

JAB DON'T CULL!

We are demonstrating today outside Maff to end the cull of healthy animals. Maff is slaughtering hundreds of thousands of disease-free animals including rare breeds, prized flocks, pedigree herds and even pet-goats. This wanton slaughter is causing cruel and intolerable human stress and suffering in addition to compromised animal welfare.

Maff has been incompetent and deceitful from the start of this outbreak. They continue to deceive as they obstruct the use of vaccination as an alternative to their environmentally appalling slaughter policies.

Maff itself must be culled

Their policy means that even if this outbreak can be brought under control, animals will remain at risk as a result of increased food transportation and movement of people and animals.

In the 21st Century, medical science has conquered polio and smallpox.

Maff remains stuck in the dark ages.

We demand an immediate end to the killing of healthy animals

Find out about Foot and Mouth Disease and how to help at

www.sheepdrove.com

The theme of a confused agricultural policy leading to both unnecessary hardships for farmers and a mass cull of animals, that could have been prevented, was complemented by increasing concern about the threat of FMD on an economy already jeopardized by recession. *Reproduced courtesy of Sheepdrove Organic Farm.*

other, quickly became a symbol both of what became known as "the countryside's anger" and also the inarticulate confusion of farmers torn between mute resignation and the desire to revolt. As the reporter Matthew Engel wrote in the *Guardian* of 17 March 2001.

> Farmers are using the same words as the TV reporters: "devastating," "terrible," "heartbreaking," "diabolical." They are used to their animals being killed . . . it is not easy for them to disentangle their true feelings. The town shares the smoke, the pain, the grief and the grievances. But the grievances are unfocussed. Is the government policy too harsh or too weak? No one is certain though they are instinctively convinced it must be one or the other. There seems to be no one in the area who can voice the farmers' feelings in a way that makes them both coherent and audible.

Britain's rural communities have never been Labour's natural constituency, and from the beginning of the foot and mouth crisis, it was clear that the Ministry of Agriculture, Fisheries, and Foods, MAFF, was not in possession of a clear policy. Not surprisingly, although paradoxically, increasing anger was expressed about the legitimacy or wisdom of MAFF's "contain-and-destroy" philosophy, at the same time that the inefficiency of its implementation was berated with equal vehemence. The *Sunday Times* editorial of 18 March called for "a return to traditional, more humane farming methods" reflecting "a national desire to rethink the farming of animals."

Under the headline "Agricultural barbarity," the *Times* lamented the "images of funeral pyres of cows, flocks of dead sheep and lambs awaiting slaughter" that had "appalled Britain and the rest of the world" by turning the British countryside into "a medieval version of hell" with "trenches filled with burning flesh and palls of black smoke rising into the sky." The foot and mouth crisis, the *Times* urged, "should make us think hard about how we treat our animals. After a week of being bombarded with television images of countryside carnage, the way we farm meat is provoking profound questions about intensive farming and the inhumane way that animals are transported around the country for slaughter." Citing a national opinion poll (NOP) specially commissioned in the wake of the government's announcement of its "action plan," the *Times* continued, "Our NOP poll today shows that four out of five people favour a return to traditional, more humane farming methods, even if this means paying more for food. More than a quarter of people plan to eat less meat as a result of foot and mouth and BSE, while 12% have already given up." Hastily interjecting that "Britain is not going to turn into a nation of vegetarians," and underlining its loyalty to the influential meat and dairy industry by adding, "nor should it," the *Times* editorial went on to criticize intensive farming methods, repeatedly calling for more humane treatment of animals, and warning of the costs of continuing "the present system." "One of the penalties for our relentless search for cheap meat has been BSE and nearly 100 human deaths from NVCJD [new variant Creutzfeldt-Jakob disease]," the editorial claimed, adding that "wanting animals to be treated better, but not wanting to pay for it is like wanting better public services and lower taxes." The solution had to be a European one, something that Britain, as "the leper of Europe on food safety," might not find "easy to champion," but should, nonetheless, pursue with all due speed, lest it bring upon itself another catastrophe of food-chain barbarism.[22]

The theme of human barbarity toward animals explicitly redefines the source of foot and mouth infection as a lack of moral values, civility, and responsibility, rather than a highly contagious virus. Terms such as "inhumane" and references to "countryside carnage" and images of "burning flesh and palls of smoke" producing "a medieval version of hell" clearly suggest a model of regression into uncivilized behavior that is both indefensible and demeaning to humans, as well as cruel and destructive toward animals. This pointed reversal of human-animal relations, whereby it is the humans who are behaving brutishly, is repeated in many accounts, such as Madeleine Bunting's outspoken commentary from the *Guardian* critiquing the irrationality and barbarity of the slaughter policy (Bunting 2001).

Under an unattractive cartoon image of grinning government ministers, soldiers, and slaughtermen taking aim at a sheep, a cow, and a pig in front of a target—set against a backdrop of a burning pyre—Bunting asks, "who are the brutes now?" Full of provocative analogies ("how can Britain criticize the Taliban for destroying the Bamiyan Bhuddas while MAFF prepares to cull Beatrix Potter's famous flock of Herdwick sheep?"), she herself takes aim at the lack of a more vigorous debate about animal cruelty. "We share up to 90% of our genetic material with these creatures who feel pain and anxiety. The most cursory philosophical knowledge exposes the flimsiness of the intellectual constructs by which we have awarded ourselves rights, and sheep and cows none," she complains. These considerations, she concludes, have "been totally subordinated to an economic rationale of productivity, efficiency and export markets from which almost no one demurs. . . . It is a terrible indictment of a culture that our accommodation with the market has so numbed us that we can see no other way of viewing the world and its occupants. Who are now the brutes when we have become so brutal?" This theme of anticommercialization, the common currency of a perennial British equation between frugality and saintliness, was echoed by none other than Prince Charles, who, like his mother, has a deep affection for the countryside and is an outspoken advocate of sustainability.[23]

As the Queen cancelled her horse show, Prince Charles took his message to Canadian farmers. The Prince of Wales, after being disinfected, expressed his well-known support for a return to more traditional farming methods to farmers in Saskatchewan at the height of public anxiety about foot and mouth. His expressed his view that his own organic farm in Highgrove could serve as a possible example of a way forward post-foot and mouth. He described the disease and

Heritage varieties of rare-breed sheep, such as the famous Lakeland Herdwicks, were rebranded as not only healthy but political in the context of what organic food suppliers, such as Prince Charles, described as the "national agony" of foot and mouth. *Reproduced courtesy of the Herdwick Meat Association.*

subsequent cull as "a national agony" and insisted that "issues surrounding sustainable agriculture must be addressed."

Following a weekend of turbulent media coverage, and less than a week after announcing the government's controversial "contiguous culling action plan," its chief scientific officer, David King, officially confirmed the epidemic to be "out of control." He called for a dramatic extension of the cull and warned that up to 50 percent of Britain's livestock might be lost. "Looking at Britain as a whole," he said addressing a press conference on Thursday, 22 March, "the situation is out of control," adding that "in the worst case scenario, out of control means that we might even lose 50% of the livestock in Great Britain" (Poulter 2001, 10). Writing in the *Financial Times*, Philip Stephens claimed that "Britain is losing its senses. The foot-and-mouth epidemic has been called a national disaster. . . . The country is showing all the signs of a collective nervous breakdown" (30 March 2001, 21). Slaughtermen struggled to cope with the physical and emotional demands of culling as many as 30 animals per day. Meanwhile, opposition to the slaughter policy become more entrenched, and groups of farmers began to threaten to take direct action, such as the woman in the Duddon Valley, quoted as the epigraph to this chapter, who claimed, "I've got a motley crew of animals. They will not get them. I've got chains, superglue, vicious geese, barricades, and I'll lock myself up with them in the kitchen" (quoted in Vidal 2001, 4).

Amid a rapidly worsening crisis, and lacking a coherent policy, the government struggled both to clamp down on militant rural protestors and to reassure the farming community of efficiency gains in the arduous culling process. On 29 March, the army began to participate in the slaughter and disposal of tens of thousands of culled livestock, increasing the sense of a national emergency and leading William Hague, the leader of the opposition, to urge Tony Blair to postpone the national election scheduled for 3 May for at least six weeks in order that the government turn its full attention to the crisis at hand. Following a tense weekend of speculation, and taking a huge political gamble, Blair confirmed his renewed commitment to an effective slaughter policy by announcing on Monday, 2 April that he would indeed shift the elections to June. Rumors about the source of the epidemic continued to spread along with the disease itself, alleging causes as diverse as smuggled meat, American tourists, and stolen viruses. Throughout early April, strict movement restrictions continued amid ongoing protests and a continuing debate about the vaccination issue.

The run-up to Easter, which fell on 15 April, saw an increasing army involvement in the slaughter and disposal processes, accelerated to a rate of twenty thousand animals per day in various parts of Britain. Brigadier General Alexander Birtwhistle compared the pursuit of FMD to combat against a new kind of enemy force, posing, he claimed, "as great a foe" as any he had encountered in his forty-two-year career. In addition to disposal problems, the knock-on effects of foot and mouth were also becoming more apparent, just as the lambing season began to reach its peak (quoted in Laville 2001, 14).

Without exception, all of Britain's Easter Sunday newspapers featured a photograph of a mud-covered lamb trapped on an exposed hillside due to the restrictions on sheep movements that prevented farmers from transporting their sheep to more protected areas for lambing. Full of resonance with Christian themes of suffering, sacrifice, and the slaughter of innocents, the lamb photo rekindled national outrage. In addition to the tenuous legitimacy of the contiguous cull policy (killing millions of healthy animals to preserve animal health), the necessity to kill ewes carrying dead lambs, and lambs that had contracted pneumonia from exposure, added painful ironies to the crisis. In late April, as the nationwide total of undisposed-of stock approached a quarter of a million possibly contagious animal carcasses, farmers reached new heights of despair in the face of MAFF's

Phoenix, the calf they couldn't cull, fights on

By David Brown
Agriculture Editor

A HEALTHY young calf called Phoenix was lucky to be alive yesterday after surviving two attempts by Ministry of Agriculture vets to kill it.

The white calf was apparently given a lethal injection six days ago along with 70 cattle — including its mother — and 47 sheep on a farm in Devon.

But when contractors arrived a few days later to spray the carcasses with disinfectant, they heard mooing and found Phoenix wandering among the dead animals. It had either survived the slaughterer's injection or been missed.

The farmer, Fred Board, and his family took in the calf, which was born on Friday April 13, and bottlefed it. When a Maff team returned yesterday to "finish the job", he defied them.

Two vets, a Maff official and two policemen left Clarence Farm, Membury, near Axminster, after a brief stand-off but warned that they would return with a court order. "It seemed a lot of officialdom for one innocent little calf," said Mr Board.

Although no foot and mouth has been confirmed on the farm the calf was caught in the contiguous cull policy to prevent spread of the disease. Mr Board said: "The calf was not injured and it appears that it was given an injection to put it down. It survived. We had the animal checked and it shows no signs of foot and mouth.

"I do not see why this healthy calf should be killed. The nearest confirmed outbreak is a mile away from me and sheep in between have been reprieved. I won't let them do it."

A Maff spokesman said last night: "Phoenix has to die. This stand will not alter what we have to do." The ministry rejected the suggestion that the calf was effectively in quarantine, indoors.

Two more suspected human cases of foot and mouth were under investigation yesterday, bringing the total to three. Paul Stamper, the carcass disposal operative from Dearham, Cumbria, who was reported with the disease on Monday, said he "wasn't feeling too bad".

Confirmed outbreaks of foot and mouth rose to 1,465 yesterday, an increase of 15, as Maff's handling of the crisis was branded "a national disgrace" by the Country Land and Business Association.

Jeremy Thorpe, the former Liberal leader, urged Tony Blair to use the Navy to dump thousands of carcasses at sea to clear the backlog in Devon.

The miraculous survival of Phoenix the calf, who survived his lethal injection and rose from the dead just before Easter, and who was subsequently given a prime ministerial pardon, rescuing him from the cull, created a nationwide celebration of the animal. *Reproduced courtesy of the* Daily Telegraph.

ongoing inefficiency.[24] Despite the army's assistance, farmers continued to report slaughtered stock being left where it lay for up to two weeks. According to the *Telegraph* of 25 April, "Linda Gifford, a dairy farmer at Holsworthy, North Devon, said her herd of 280 cattle, slaughtered on April 10th, were still lying outside her kitchen window more than two weeks later." The bizarre and rather gruesome incident of a laborer who, in an attempt to move a bloated carcass, caused it to explode and involuntarily ingested some of its contents, led to the first case of human infection with foot and mouth—a rare and non-life-threatening event—but one that inevitably fueled well-worn national anxieties about human-animal connections as a source of contagious transmissions. Offsetting such unpleasantness, and perfectly timed as an antidote to the mud-covered lamb debacle, was the rescue and "pardon" of Phoenix, the calf.

Born on Good Friday (ironically the thirteenth), Phoenix was given a lethal injection two days later, along with his mother, seventy other cattle, and forty-seven sheep—only to be found alive and well the following week, wandering amid the corpses of the other animals, by farmer Fred Board. Board named the youngster Phoenix because of his spectacular rise from the dead and adopted the animal as a family pet. In the wake of his family's grief at the loss of their livestock, it seemed excessively harsh to cull the perfectly healthy young Phoenix. Protested Board, "I do not see why this healthy calf should be killed. The nearest confirmed outbreak is a mile away from me and sheep in between have been reprieved. I won't let them do it" (quoted in Brown 2001, 1). Overriding MAFF's insistence that the animal must, despite its good fortune to escape the cull, be put down, Tony Blair responded to Phoenix's newfound national celebrity status as a heroic survivor by granting the animal an official prime ministerial pardon.

Despite the fact that the crisis continued to worsen throughout May—and in the teeth of ongoing, and as yet unresolved, debates about vaccination—Blair held and won the national elections in June—by a landslide. William Hague immediately announced his decision to step down as leader of the Tory party, and was replaced by Ian Duncan Smith. In the previous three months, nearly 4 million of Britain's livestock had been killed, approximately 7 percent of the country's estimated 62 million farm animals.

After Foot and Mouth

In the wake of foot and mouth many factors have been identified in explanations of how the epidemic took hold and spread so quickly, although the value of such post hoc analyses remains somewhat hazy. For example, it was the highly mobile system of sheep transport, whereby animals are carted up and down the country in a kind of calculus of sheep exchanges, which both launched and enabled the rapid spread of the disease, thus causing the epidemic. However, this stratified system of crossbreeding has been at the heart of the British sheep-farming industry for centuries, and it is unlikely to change. The question of vaccination, so vexed throughout the controversy, is also unlikely ever to be solved. Vaccinations only work for specific strains of FMD, of which there are many, and they are opposed by the dairy industry, which fears negative customer perceptions of the resulting milk and milk products. Neither is the question of whether MAFF's policy was too strict or not strict enough likely to ever yield a definitive yes or no answer. It may be, as some expert commentators have suggested, that a more rapid involvement of the army would have saved many sheep's lives. However, such a view could easily lead to the very overreaction in the next outbreak that many argued characterized the most recent one.

Concerns about infected meat from Chinese restaurants, pigswill containing airline food, and the spread of airborne BSE prions are just the kinds of new biorisks that take shape and become visible because they are seen to introduce new vectors of contagion. Indeed, foot and mouth has occasioned a number of significant realignments in the organization of rural lives, economies, and environments. It has both reinforced and changed the traditional meanings of the rural, the countryside, and agriculture. Above all, the somewhat surprising message to emerge out of FMD is that it is because the countryside is now more modern and industrialized that it must incorporate more traditional elements. The ideal politics of the food chain are increasingly expressed in terms of balance: the benefits of modern industrialized agriculture must be balanced against some of the lost features of more traditional farming.

In sum, what foot and mouth reveals are the two sides of biological control. On the one hand, modern industrial sheep farming involves an unprecedented level of control over the animal, from the genetics of its breeding to the composition of its diet and the ability to control its movements. In every aspect of sheep farming, from shearing to transport, important advances have been made that benefit both the sheep and the farmer. However, while this level of control is cumulative in some of its aspects, it is irrelevant to other dimensions of sheep farming, and may even constitute a contributing factor to problems such as foot and mouth—in part brought about by the high-speed transit of live animals. In the face of foot and mouth—one of the oldest of sheep diseases, and only lethal because markets are global—the most important technologies are the oldest, simplest, and most reliable. The scorched-earth policy relied exclusively on culling and burning while movement restrictions relied on path and road closures. While Dolly remained quarantined in her Scottish paddock, embodying the most sophisticated technologies ever used to create a sheep, all around her an ancient disease was fought with ancient methods, confirming that the acquisition of ever greater biological control has no necessary relation to the reduction of its opposite. Like

The Guardian Saturday May 26 2001

Foot and mouth crisis

Milk may contain dioxins from animal pyres

Britain's FMD epidemic of 2001 was the occasion for what some have described as a seismic shift in its attitudes to food, farming, and the countryside. Ongoing debate about the management of the epidemic has led in turn to greater awareness of the intimate links between the countryside, farming, food and health. *Reproduced courtesy of the* Guardian.

Concern about the survival during FMD of one of the nation's best-loved breeds of sheep, the Herdwick, illustrated the extent to which these animals continue to embody national heritage—in this case both because they are linked to the Lake District, and were bred by Beatrix Potter. *Cumbria Life*, April 2001.

increased security, increased control produces new risks and may aggravate susceptibility to older ones.

The consequences of foot and mouth can at one level be summarized as a concern with the relationship between industrialization and domestication. It is widely felt that the overdomestication of animals, plants, and soil produces degeneration—perhaps in the sense of being overrefined so much that they become once again crude.[25] The in-and-in breeding methods used to produce purebred types—the first truly modern breeding system developed by Robert Bakewell—shares some of this paradoxical quality. Foot and mouth posed the paradox of domestication in a wider, more systematic, and arguably more global way, and in ways that brought it home in a violent and, some claimed, barbaric manner.

This paradox, like debates about cloning and Dolly, the sheep, illuminates how anxieties about animals both replicate and substitute for human concerns about their own reproductive futures. In terms of fertility, subsistence, sustainability, and, most broadly, of health and its connections to animals and soil, the food chain now raises a host of new kinds of political issues that are both intimate and emotional, as well as local and quotidian. Animals thus increasingly embody the local/global politics of food and health production, and they bring these economies home. Returning to the image with which I began—the somewhat unexpectedly heroic image of sheep—it is clear that the global rural remains inextricably bound up with the rural animal, whose futures appear ever more closely tied to our own (Franklin 1997b, 2001c).

Dead Sheep

The ability of the rural farm animal to evoke what Raymond Williams calls "whole ways of life and whole ways of struggle" captures one of the most important sources of emotional attachment to sheep as embodiments of human labor, industry, and accomplishment. Similarly, the public grief of Australian Prime Minister John Howard over the rejected Australian sheep at sea demonstrates their ongoing importance to a sense of national identity. Dolly embodied these capacities too. As we saw in chapter 3, she formed part of a sheep triptych proposed to rebrand Britain, and in the same year her likeness was proposed as a fitting sculpture to top the empty third column in Trafalgar Square. Endlessly featured in magazine articles, news stories, and cartoons, Dolly became a livestock animal representing the future of human health, reproduction, and scientific achievement. As one Roslin stockman said to me, "she is talked about by taxi drivers in China." At Roslin, too, Dolly was an iconic animal "at home"—a part of the family despite her celebrity. Frequently pictured with members of the scientific community at Roslin, in particular with Ian Wilmut, Dolly very much formed part of the team—put out to pasture like a champion racehorse to live out her days in peace.

Her days were fewer than anticipated and as noted at the outset of this chapter she was euthanized in February of 2003 after being diagnosed with a common infectious and progressive lung disease she likely caught from another sheep in her paddock. Following earlier reports of her premature arthritis and speculation about the length of her telomeres, which possibly indicated she was "born old," newspapers quickly associated Dolly's death with the known pathological consequences of cloning.[26] Not long after Dolly's death, in July 2004, PPL Therapeutics, Scotland's leading biotechnology company that had provided the cells for her birth, announced its plans for liquidation, following the collapse of its major research initiative involving the enzyme AAT. These two demises—of Dolly and of her parent company—leave many questions in their wake about both the future of transgenesis in farm animals, as well as the application of somatic cell nuclear transfer to medicine. They reveal the uncertainties, both scientific and commercial, that beset the biotechnology industry—for which Dolly was a poster sheep, but may be remembered as a prematurely aged celebrity.

The science of cloning has flourished in Dolly's wake. Since her birth, the basic principles of the Dolly technique have been confirmed through a series of related experiments on sheep, cattle, mice, rabbits, goats, pigs, and cats. In 2005 Roslin was granted the first license in Britain to use the Dolly technique on human eggs, to create human embryonic cell lines, with funding from the institute's partner company Geron.

Like the paths of sheep before her, Dolly's genealogy now leads everywhere. In the spring of 2004, the former U.S. first lady Nancy Reagan testified before the Senate in favor of reversing George Bush's ban on stem cell research, so that the benefits of the Dolly technique could more quickly reach the stage of therapeutic application. Meanwhile, in Britain, in the wake of foot and mouth, the Rare Breed Society has inaugurated a biobank to preserve the bloodlines of carefully tended flocks of heritage sheep, such as the Lakeland Herdwicks. In the future, as imagined by some, somatic cell nuclear transfer may be used to resurrect extinct mammals, such as the Tasmanian tiger, and to ensure the preservation of endangered species.

In all of these ways, Dolly's legacy of immortality could be seen to exceed the events of her brief existence. Fittingly, she now stands in Scotland's Royal Museum as a taxidermied tribute to her national and international importance. Assembled by one of the world's foremost departments of taxidermy, she has been eviscerated, cured, recast, and reassembled to acquire the desired appearance of reanimation, mid-gesture—in this case turning her head slightly to the side in the manner characteristic of how she greeted her guests.

However, it is unlikely that the legacy of Dolly's death will be quite so easily surpassed by her many successor clones, or clonal lineages of stem cells, human or otherwise. The promise of the technique used to create her notwithstanding, Dolly's creation demonstrates the narrowness of the line between life and death—a point her early demise now underscores.

A more likely Dolly legacy will be the opportunity to reexamine her life as an important window into the changing relationalities connecting humans and other animals through newfound technologies of reproduction binding them together into new assisted genealogies that are partially connected and potentially remixed. In this context, the remaking of genealogy will pose a challenge to the imagination and to description much in the same way that the populations of human embryonic cell lines banked for future use will. It is possibly the question of what kind of sociality these connections belong to that Dolly helps us to ask.

Breeds

> We have then always to be prepared to speak of production and reproduction, rather than of reproduction alone. Even when we have given full weight to all that can be reasonably described as replication, in cultural as in more general social activities, and when we have acknowledged the systematic reproduction of certain deep forms, we have still to insist that social orders and cultural orders must be seen as being actively made: actively and continuously, or they may quite quickly break down. That some of this making is reproduction, in its narrowest as well as in its broadest sense, is not in doubt. But unless there is also production and innovation, most orders are at risk, and in the case of certain orders (most evidently that of the bourgeois epoch centred on the drives of capitalist accumulation) at total risk. Thus, significant innovations may not only be compatible with a received social and cultural order; they may, in the very process of modifying it, be the necessary conditions of its reproduction.
> —Raymond Williams, "Production and Reproduction"

Dolly is a new breed, born from an old form of breeding, crossed with new forms of reproduction, as a form of innovation. The Dolly technique, imagined as an application for transgenesis, became the base, principal technique, or lifestock, of the most significant area of postgenomic research investment, namely, stem cell science and tissue engineering. Dolly consequently belongs to a familiar descent pattern of agricultural experimentation in pursuit of new forms of technological assistance to reproduction in order to generate products for new commercial markets. She is a designer animal in a bioeconomy increasingly organized around the bespoke component, the engineered pathway, and the biomimetic assemblage. Thus, in thinking through what kind of breed Dolly turns out to be, and how her breeding is linked to other genealogies, including our own, this conclusion begins by thinking about what we might call "smart sheep."

In November 2001, Keith Kendrick, a behavioral scientist, and his

colleagues from the "sheep team" at the aptly named Babraham Institute at the University of Cambridge published the findings of a lengthy study in the journal *Nature*, indicating that sheep have highly developed capacities for facial recognition and memory (Kendrick et al. 2001). Using twenty-five pairs of photographs of sheep faces, one of which was associated with a food reward, over a series of thirty trials, Kendrick and his team demonstrated that sheep could remember up to fifty faces with 80 percent accuracy. The Babraham sheep were, in addition, able to identify unfamiliar profile shots of sheep they had only encountered face-on, and to retain this information for two to three years. Preliminary neurological findings also suggested that the sheep participating in the study could form a mental image of an absent sheep, leading some media commentators to suggest that while sheep appear to be in "a grazing stupor," they may be ruminating on more complex matters, or even picturing their lost companion animals: "As they stand huddled with the rest of their flock in what appears like a grazing stupor, sheep may in fact be visualising long departed flock-mates" (Trivedi 2001, 1). Predictably, Kendrick's study elicited significant media coverage of smart sheep: the headlines included "Amazing Powers of Sheep" (BBC News), "Sheep Can Remember Faces Says Professor" (BBC Wales), and "The 'Intelligent' Side of Sheep" (BBC Wales).

Sheep intelligence, and the animals' highly developed ability to recognize other sheep, was keyed to their sociality, endowing the much-ridiculed sheep brain with greater dignity and pathos, as well as complexity and sophistication. As Kendrick revealed to Steve Connor, the science editor of the *Independent* newspaper, "Sheep form individual friendships with one another, which may last a few weeks. It's possible they may think about a face even when its not there" (qtd. in Connor 2001). Kendrick's suggestion that sheep can "respond emotionally to individuals in their absence" was interpreted by animal welfare organizations, such as People for the Ethical Treatment of Animals (PETA), as evidence that "sheep have feelings too" (PETA 2001).

Sheep Society

As Thelma Rowell, a leading primatologist turned sheep observer, has argued, perceptions of sheep reflect the eyes of their beholders, and they are often as anthropomorphic as they are self-fulfilling. The view, for example, that primates have the most sophisticated social relationships of the higher vertebrates may well be an artifact of the expectation that they most resemble humans, a view reflected in the tendency for research on primates to involve "very long term studies in which the social interactions of known individuals are followed for long periods," focusing on their "familial relationships." This methodological tendency may itself be self-perpetuating in the sense that "perhaps since long-term social effects are expected, they have often been found" (Rowell and Rowell 1993, 214).

Based on her extensive ability to compare the sociality of sheep and primates, Rowell argues that a capacity for ovine sociality could be demonstrated not only to equal but to surpass that seen in non-human primates. Sheep, she argues, not only have "the capacity for maintaining long-term individual relationships among adults" but also the "basic requirement for building society" (Rowell and Rowell 1993, 228).[1] Nor is this sociality defined in exclusively hierarchical terms, as some sheep ethnologists have previously suggested (Geist 1971), but is, rather, expressed through a wide range of behaviors, by both rams and ewes of differing ages, including "affiliative reassurance behaviour" (head rubbing), which suggests "a level of sophistication [that] has not yet been described among non-human primates" (Rowell and Rowell 1993, 230). As Rowell also observed in her and her daughter's study of feral sheep behavior,[2] "third party intervention in agonistic interactions between pairs also suggested appreciation of a wider social structure rather than a system of one-to-one responses" (1993, 230). Such evidence of sheep sociality is of intrinsic interest in its own right as a contribution to the understanding of social structure—be it among ruminants, primates, or humans, or some combination of these and other animals. However, what is also of interest is the suggestion of anthropomorphism implied by the tendency mentioned by Rowell of researchers to identify with other primates, triggering its corresponding, self-fulfilling, impetus. As Rowell and Rowell point out, sheep are an ideal subject to test preconceptions of animal sociality, its sophistication, or the absence thereof, precisely because domestic sheep are "popularly taken as the very paradigm of both gregariousness and silliness" (1993, 214). As we have seen, sheep are also often viewed as stupid, simple, mindless, and timid—and have thus become the subject of endless supposedly primitive humor.[3]

For Rowell, as for many anthropologists, the concept of sociality is inseparable from the methods used to study it. For example, in her review of Dale Lott's 1991 comparative synthesis of nonprimate vertebrate sociality, *Intraspecific Variation in the Social Systems of Wild Vertebrates*, Rowell asks whether it is useful to approach the question of animal sociality through the model of a "social system," which implies fixed rules of conduct: "If differences in the behaviour of populations can be described in terms of differences in the behaviour of individuals, what is gained by speaking in terms of a social system? What is a social system anyway? *Is it possible that the social system is a concept that might actually get in the way of understanding social behaviour?*" (1993, 135; emphasis added). Rowell describes this problem as the "reification of the social system," which she opposes to the attempt to identify ranges of variation in both individual and group behaviors over time. This also leads to a question of perception: Are social systems sets of rules that determine behavior, and thus perceived by animals themselves? Or is the idea of a social system something that can be seen as an emergent property of the long-term behavior of animals in stable groups, something that is only visible to the observer as a retrospective composite of these behaviors? (Rowell 1993).

Sheep are, for all of the reasons this book suggests, a useful animal model with which to consider the structural, functional, affective, and performative dimensions of human, as well as animal, sociality. In addition to being animals who move in between feral and domesticated lifeways, they are also unusually highly imbricated into advanced human sociality by virtue of their crucial importance to human settlement, migration, industrialization, agriculture, industry, religion, nation formation, and, increasingly, health care and biomedicine. Can we extend Rowell's question of sociality to include the range of sheep-human behaviors that gave shape, for example, to the type of Australian sociality—argued to be at the core of the Australian "character"—that quickened into existence out of frontier settler pastoralism? Are the ways in which sheep and human behaviors structurally linked—for example, through the wool trade, frontier settlement, or Scottish hill farming—examples of innovative, enduring, and distinctive relationalities that give a richer texture to the vicissitudes of domestication, revealing, as in the case of pastoralism, the extent to which its ironies, tensions, and contradictions indeed constitute its enabling condition?

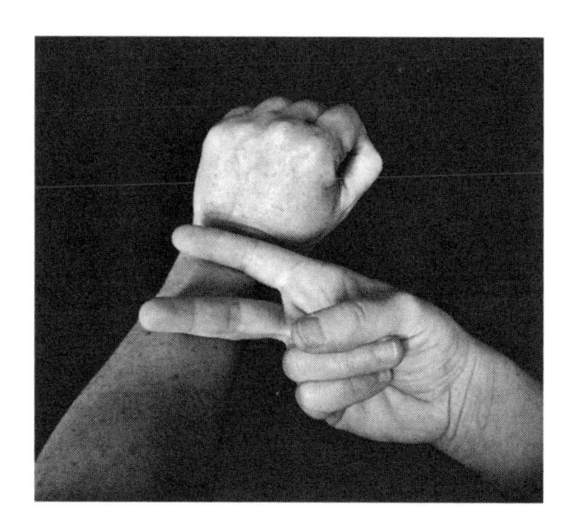

The downward clipping movement in the two-handed sign for "sheep" performs a dense manual conflation of the animal, the tool used to clip it, the action of the clipping, and the product of this exercise, which is wool. Significantly, then, the sign performs "sheep" as a human/animal/technology/commodity mixture. *Photograph by Sarah Franklin.*

Sheep Signs

Consider for a moment the hand movement in American Sign Language (ASL) for *sheep*, a downward clipping movement with two fingers against the opposite forearm, imitating shearing.

Through this digital miming gesture, the animal is represented by the shearing process, and the clipping movement enrolls the signer's body into a conflation of animal, technological, and human agencies that literalizes the union of all three as a word. In the signing action we find a number of revealing semiotic condensations: of the sheep with its wool, of its wool with shearing, of shearing with a handheld device, of the animal with its human utility, and of the sheep's valuable fleece with the manual action used to remove it. The hand language for sheep is thus doubly manual: a manual imitation of a manual task. Above all, the sheep emerges as something that is held, its body represented not by the hand, but by the digitless, passive back of the forearm, against which the manual shearing/signing action is performed. One hand clipping signs *sheep*.

This signing action thus performs and reinforces the widespread perception of sheep as essentially passive, lucrative, and tractable (a view few shearers share), through which in turn they are associated with meekness, timidity, and stupidity. Yet this view switches, or retracks, when sheep are celebrated as the herd animal par excellence whose successful introduction into environments in which no other animal, or even humans, could survive has made of their stoic perseverance and suffering an icon of rural adversity, frontier hardship, economic prosperity, and even religious salvation. What could be described as digital about Dolly's biology is similarly her ability to track back and forth between her "other" status as a sheep and her "us" status as a medical model—a well-trodden path in the wake of IVF, where this connection was first made explicit in the context of human reproduction. More than any other domestic animal, the sheep so closely resembles humans that it has long been considered an ideal substitute for humans in medical scientific experimentation, particularly for questions related to reproduction and respiration.

When I began research on sheep, I was intrigued to learn more about the attribution of stupidity to these animals and frequently asked breeders about this seemingly universal opinion. In my overactive ethnographic anticipation, I speculated that the general sheep stereotype might differ significantly from the views of those who worked with sheep on a daily basis—surely sheep breeders must at least think some sheep are smarter than others, I reasoned. Thus expecting some insider respect for sheep, I was struck by the almost vehement tenacity of the sheep-are-stupid view, even, and to my surprise sometimes *especially*, among seasoned sheep breeders. Proof of ovine mental deficiency frequently included the claim that sheep were prone to particularly "stupid" ways of dying, often by falling head first into places or positions from which they could not extract themselves, or catching their head on obstacles such as fence rails, and then breaking their own necks trying to escape. I heard endless stories of stupid sheep deaths.

As I was making sense of these sheep denigrations, I met an anthropologist from China, Yunxiang Yan, who had formerly been a shepherd himself.[4] He interpreted the stupid view of sheep as an artifact of the Western tendency to equate individualism with intelligence, originality, and leadership. In China, Yan pointed out, the association was the other way around. "Anyone can be an 'individual,'" he pointed out dryly, the implication being that it took little talent to be iconoclastic or to stand out. In China, he went on to explain, where conformity is a competitive social skill and the point is precisely *not* to stand out, the sheep is considered a highly intelligent animal.

Reading Dolly

In many readings of Dolly, it is her sheepish sameness that makes her a nonthreatening, or even comic, animal. The frequent use of humor in response to Dolly, and in particular the frequency of punning, or double entendre, reflects her doubled, or mixed, significance. At one level, her substitutability for the human creates anxiety, but this is offset by the reassurance that she is the most tractable of animals, epitomizing human control. Hence while the dangerous human-*as*-animal potential she signifies is offset by the reassuring human-*versus*-animal history sheep embody, this again is confused by the human-*via*-animal situation Dolly intensifies.

This paradox of domestication, mentioned in relation to Harriet Ritvo's work earlier (1987, 1997), and increasingly the object of anthropological reassessment (Cassidy and Mullin forthcoming), is the origin of a complex affective economy only hinted at in the previous chapter. As the Australian poet John Kinsella writes in his poem "The Epistemology of Sheep,"

> Ah, the tame ones, privileged
> In the hierarchy of sheep—
> Uggboots of the conscience,
> The cosy lambwool
> Cardigan, the tough
> But comfortable carpet.
> (2001, 49)

Here, the domestic animal is represented as moving in and out of leaky categories: inside and outside, tame and wild, privilege and abjection, and above and below through everyday objects made of sheep products—Ugg boots made of fleece, cardigans made of lambs wool, and carpet made of the thicker, shorter wool fibres of purpose-bred sheep types. It is the proximity and everyday utility of these objects, at once cosy and tough, that tug at the human conscience in their inescapable reminder of the animal death that sustains human life.

Too Clone for Comfort

In her influential analysis of what she calls the "ontological choreography of new reproductive technologies," feminist science studies scholar Charis Thompson argues, following Marilyn Strathern (1992a, 1992b), that the increasing ability to "assist" human reproduction has led to "a deftly balanced coming together of things that are generally considered parts of different ontological orders," such as nature and society, biology and technology, the private and the public, or self and other. The assisted conception clinic, she argues, stands as a particularly salient example of the processes whereby "these elements have to be coordinated in highly staged ways so as to get on with the task at hand" (2005, 8), by normalizing, renaturalizing, or realigning messy hybrid components into their proper shape. This process, she argues, is how parenthood is now made, much as it has been described as a new form of "achievement" (Franklin 1997a), "accomplishment" (Ginsburg 1989), or part of the work of consumer culture (Layne 1999; Taylor et al. 2004).

For example, the process Thompson describes as "strategic naturalization" in the context of assisted conception allows for a non-traditional means of impregnation, such as conception with donor sperm, to be deliberately realigned with more conventional norms of descent through a selective emphasis on some natural facts (such as maternity via pregnancy) rather than others (such as conception via donor). Other authors, including Helena Ragoné (1994) and Corinne Hayden (1995), have made similar points about how kinship is "naturalised, denaturalised, and renaturalised" in the context of assisted conception (Franklin, Lury, and Stacey 2000).

The mobility of naturalization, denaturalization, and renaturalization processes is similarly emphasized by observers of science in the making, such as Bruno Latour, who describes the networks of action though which one thing can become another and through which ontologies are stabilized as nouns—as when an unknown substance from the hypothalamus is experimentally demonstrated to inhibit the release of growth hormones and thus becomes an entity: somatostatin (1987, 88).

The ontological choreography surrounding the life and death of Dolly evokes the same mixture of comfort and discomfort described by the poet John Kinsella in his account of the connections and dis-

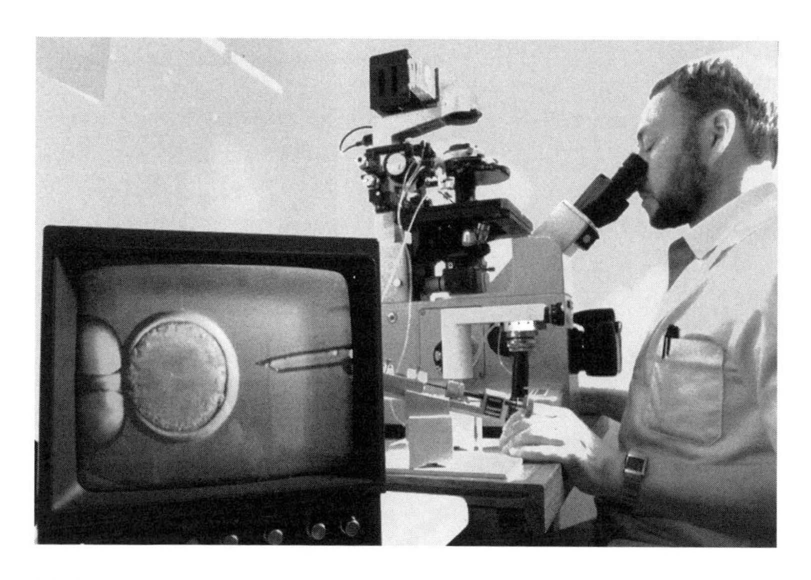

Nuclear transfer in progress. The natural interface between micromanipulation and the ability to screen the embryo has resulted in a proliferation of microinjection imagery, such as that used to create Dolly, as demonstrated here by Roslin's Bill Ritchie. Questions about the effect of cloning on familiar economies of similarity and difference have become particularly evident in popular visual culture, where imagery of micromanipulation has come to play an increasing role, signalling both a new primal scene of life's creation, but also the threat of loss of sex as an origin and guarantor of individuality and difference. *Courtesy of Roslin Institute.*

connections that "have to be coordinated in highly staged ways so as to get on with the task at hand" (Thompson 2005, 8). As Arlene Judith Klotzko observes, the figure of the clone is "deeply threatening to our sense of individuality and autonomy," which constitute "crucial values, especially in Western societies" (2004, 150). Thus cloning is associated with what Jackie Stacey, in her analysis of the new genetics and Hollywood cinema, describes as the "fear of sameness," a fear she links in particular to the "horror" of same-sex reproduction (Stacey 2005). This horror of illegitimate copying is reproduced within the capitalist economy in a prohibition on cloned products, seen as illegitimate in the same way "bastard" offspring are when they are shamed by a lack of proper parentage. The same horror of duplication and taint of fraud stigmatize the figure of the double, replicant, or multiple in ways that are both gendered and racialized, as the anthropologist Deborra Battaglia illustrates in her 2001 analysis of popular Hollywood cloning narratives.

The fear that attaches to the figure of the double, clone, or copy is thus not only of the loss of originality as identity, a form of individual uniqueness that affirms the superiority of the two-into-one logic of sexual reproduction and bilateral descent. It is also the fear of the asymmetry, or difference, between the original and its second, which is most often expressed as a fear of inequality, inferiority, and vulnerability to being used for another's purposes, or appropriated to another's ends. The fear of being made to order or copied is one of loss, devaluation, and worthlessness—of being derivative, barren, and unoriginal.

Sheep in service as domestic livestock animals live in almost exclusively female groups and are frequently in-bred to a degree that leads not only to high levels of physical resemblance (to a nonsheep) but also of genetic similarity, so much so that purebred animals are considered virtual clones of one another. While seconding humans and embodying each other as multiples in a highly feminized social order, sheep also, and this is perhaps no surprise, elicit paradoxical emotions that require strict rules of proximity, distance, and affect to be stage-managed. To have sheep's eyes is to be besotted, to be sheepish is to be bashful, and to separate the sheep from the goats is to extract what is most valuable from the rest.

The invisible differences of sheep form, of course, part of the paradox of these supposedly silly animals who commonly represent the hapless victims of fraud or the passive recipients of injustice, while at the same time being admired for their stoicism and even worshipped for their sacrifice. The intelligence of sheep, and their social sophistication—obvious features of their exceptional ability to survive—are masked by their apparent tractability and collective flight response to perceived aggression in ways that service the desires of their keepers to keep them mentally in place—a struggle that can all too easily be disrupted by the silence of the lambs, as became so overwhelmingly evident in the context of the British foot and mouth epidemic. Sheep can be extremely aggressive toward perceived intruders, and both ewes and rams are known to joust as part of their routine social interaction. Like dogs, sheep will readily bond into close relationships with humans, by whom they are frequently adopted as pets, in which context, as was the case with Dolly, they become more recognizably canine in their devious and imploring obsessions with food and being fed.

Reading Dolly's mixtures thus requires us to consider how these contradictory features of sheepishness interconnect. Why, if they are highly intelligent and socially sophisticated herd animals that are more fully integrated into human history, economy, religion, migration, settlement, and identity than almost any other animal (except the dog), are sheep regarded with such abject disregard, and is there a relationship between these two aspects of sheep perception? Is it *because* the sheep is a human double that it is so abjected? Or is it because they are so abjected that sheep are so readily bound up in human sacrificial rites? What is the abject object of sheepishness? Is it the animals' secondment to human ends? Are their genealogies shameful because they implicate the dispossession of domestication? Do sheep genealogies not already haunt the human imagination with the can-be-done of made-to-order purebred lines culled to a fine point of human purpose in the name of agricultural and industrial progress? Are not the careful forms of border control used to organize sheep lives and sheep deaths, as well as sheep crosses and sheep births, the parameters of a genealogical grid that also maps the unknown future of human germinal ambition? Is it not inevitable that we become reverse-engineered in this scenario? Are stem cells a vector through which domestication is being extended—through enclosure, cultivation, and improvement to the frontiers of our own interiority?

Double-Ewe, Double-Ewe, Double-Ewe

The horror of cloning as sameness, loss of identity, and diffusion into a woolly mass may be readily captured by analogy to an indifferent flock of look-alikes, but since sheep also mirror the human condition that they, more than any other animal, have helped to shape, this question might be said to have a troublesome supplement. Perhaps by the logic of this supplement Dolly *was already a human clone* when she was born. Such a perspective, which resituates the cloning question as one of the human authorship of genealogy, is perhaps returned as the colonization of human interiority that sheep's clothing can only partially mask.

Like the unusual twenty-third letter of the English alphabet, the double-u, now routinely trebled as a shorthand for the Internet, Dolly

Morag was the first sheep to be taxidermied at Scotland's Royal Museum, where she now stands with Dolly. Recreated as an imitation of her live self, Morag and the cleaner also suggest an urban imitation of a rural scene. Set in the middle of the vast marble floor of the museum, the scene is equally suggestive of agricultural, institutional, and domestic space. *Reproduced courtesy of Murdo Macleod.*

constitutes a novel and distinctive supplement to the social, biological, and economic orders that preceded her—orders her birth both *confirmed* and *transformed* through duplication. Her sheepishness may be arbitrary in so far as cats and mice, cows and pigs, and other animals have all been cloned since she was (and some in different ways before her). Nonetheless, her ovine ancestry seems apt for thinking about what it is, exactly, that she asks of her human audiences, with whom her double association is both as a clone and a sheep. Partnering the fear of genealogical dispossession she doubly stands for is the more subaltern realization that she perhaps awakens: that genealogical propriety itself constitutes an imitation of life's imagined natural order. This natural, or biological, order will not provide the genealogical reckoning necessary to chart the future, any more than it was the only one to guide the past. It may thus be the shared genealogy of remaking that we and Dolly both belong to that establishes a truer frame of reference for asking what it means to cultivate the future of the human. It is likely to be a future in which we continue to have more in common than we think with sheep. This is, of course, the reason why the remaking of Dolly's genealogy matters—because it remakes our own.

Notes

Origins

1 An excellent account of Dolly has been written by the *New York Times* science journalist Gina Kolata (1997), which includes a helpful introduction to the history of the science of cloning. The most comprehensive account of the making of Dolly, on which this book relies substantially, is that written by the Roslin scientists Ian Wilmut and Keith Campbell together with the British science journalist Colin Tudge (2000). The bioethicist Arlene Klotzko's edited sourcebook on cloning is invaluable (2001). Klotzko's more recent monograph on cloning (2004), like that of the philosopher John Harris (2004), is primarily concerned with the ethical implications of cloning. Andrea Bonnicksen's discussion of the public policy dimensions of cloning and stem cells (2002), although primarily based on the United States, is extremely helpful. For a review of cloning literature, see Franklin 1999b.

2 On the meanings of "procreation" in the context of assisted conception, see in particular Strathern 1995, 1999b. See also chapter 1 of the current volume, which concerns this topic.

3 For social and historical analyses of the importance of control over programmed cell death in the production of cellular value, see Landecker 2000, 2003; Lock 2001; Hogle 2003; and Squier 2000. For a polemical account of the science and politics of stem cells in the United States, see West 2003, and for an indication of the increasing importance of cell immortality within popular culture, see Amazon.com, where in March 2004 a search of that phrase produced 14,601 hits—many from the new age and lifestyle literatures.

4 For an analysis of stem cells as mixtures, see Franklin and Lock 2003a, 2003b; Franklin 2001b, 2003b, 2005. For their connections to social movements, see Rapp 2003.

5 For a detailed account of Britain's agrarian imperialism, see Drayton 2000; Fara 2003; and Gascoigne 1994, 1998. For contrasting accounts of ecological imperialism, see Anker 2001 and Crosby 2004. See also Haraway 1988 and Mackenzie 1988 for accounts of how U.S. American and British imperial accounts of nature both emerged from, and contributed to, wider projects of technological innovation, political economy, and national identity formation.

6 Dolly is, in this sense, what the Australian postcolonial theorist Nicholas Thomas describes as an "entangled object" (1997).
7 For an excellent review of the "animal turn," see Mullin 1999. See also Franklin and White 2001; Martin 1995; Willis 1990; and Wolch and Emel 1998.
8 For further discussion of domestication, see Anderson 1997; Cassidy and Mullin 2007, Clutton-Brock 1987, 1989; Ingold 1988; and Leach 2003.
9 My view of genealogy, in the strict sense of the term (meaning ties established through shared reproductive substance, that is, through blood or DNA, which are the ground zero of genealogical reckoning), is that, like sheep, it leads in unpredictable directions and "gets everywhere." This argument is beautifully illustrated in the work of Catherine Nash, who shows how even when people go in search of "genetic" genealogy, they often find unexpected, nongenetic, and just as affective bonding experiences (2002, 2004). The nonliteralism of the new genetics, for example in the context of prenatal screening, is also an example of what might be called the degeneticization of genealogy in exactly the place one might expect to find the reverse (Rapp 1999; also see Edwards 2000; and Franklin 2003d).

1. Sex

1 Biologically, *replication* is defined in the *Oxford English Dictionary* (OED) as "the process by which genetic material or a living organism gives rise to a copy of itself," as in cell division, whereas *reproduction* is specifically associated with the "process of producing new individuals." *Regeneration* is distinguished by its reference to the process of "reproducing parts of the organism which have been destroyed or removed," as in the ability of tissue to regenerate.
2 Individual organisms combine several types of reproduction naturally, in the sense that body cells are reproduced through asexual division or mitosis (preserving their genetic identity), sex cells are reproduced through recombination by meiosis (creating new genetic mixtures), and stem cells are somewhere in between, producing both identical "daughter" cells that remain progenitor cells and other daughter cells that differentiate to become specialized cells, such as blood or bone marrow cells.
3 Transgenic sheep containing desired genetic modifications are described as bioreactors, designating their ability to perform specific kinds of manufacture, such as producing proteins in their milk. Dolly, who is not transgenic, could be thought of as a nuclear breeder reactor in that she, too, is a model of a specific kind of manufacture—not through her milk, but

in her embodiment of the viability of nuclear transfer using differentiated adult cells from culture.

4 A significant feature of early-twenty-first-century biotechnological discourse evident in Wilmut's descriptions of Dolly's creation is the switching back and forth, relay, or transfer between idioms of propagation, cultivation, and growing, on the one hand, and engineering, design, and building, on the other. This is an example of what Marilyn Strathern (1992a) refers to as a "merographic connection," in that parts of these idioms overlap, while as wholes their meanings exceed each other. As we shall see in the next chapter, the core symbol, or condensed node, of these intersecting idioms is the idea of stock.

5 The expression *proper clone* is frequently used in relation to Dolly, as in the assertion that she is not a "proper clone." This leads to the question of what a "proper" clone would be, and what is improper about the kind of clone Dolly is. In turn, this contrast reveals the very general and imprecise nature of the term *cloning*, which allows it to signify so broadly, and perhaps so threateningly, because of its semi-feral semantic wanderings.

6 Her shared genetic identity is both shared and an identity in two simultaneously opposite senses, thus again performing the Dolly paradox. Her genetic identity is not strictly her own because it is *identical* to another, with whom she shares it both in the sense of having "part of it" (a share) and "all of it" (they share the same *complete* identity).

7 For a detailed history of the scandals associated with cloning, see Kolata 1997, especially chapter 5; and Wilmut, Campbell, and Tudge 2000, part 2. See also Jon Turney's (1998) detailed study of Frankenstein mythology in popular accounts of cloning.

8 In her pathbreaking work on new genetics and the cinema, the film theorist Jackie Stacey (2005) argues that the clone belongs to an "economy of sameness" evoking horror in films such as *Alien: Resurrection* (dir. Jean-Pierre Jeunet, 1997) because of its severed ties from heterosexual kinship and its consequent lack of what might be called proprio-filiation, or genealogical alignment (see also my final chapter here).

9 I am using *queer* here in the sense of both "unusual" and "sexually deviant," since Dolly's genealogy is both. It is, in addition, not "straight," either sexually (i.e., reproductively, as she was not produced through straight sex), or in terms of descent (her genealogy is not properly aligned; see further in Ahmed 2006). Interestingly, though, her birth is associated with truth, in the sense that her viability confirmed new facts about life's continuity, disproving former beliefs about biology. By definition, these new facts of life are counternormative.

10 In asking if we "have ever been modern," and concluding that we only have to the extent that we successfully conspired in the impermeable categories that made us so (e.g., nature), the science studies scholar Bruno Latour (1993) suggests a reversed version of the question that could be asked after Dolly of whether it is only through her, and her kind, that we truly ever have "become biological."

11 Dolly's troubling identity as a clone is also hidden by the fact that she is a sheep, who are imagined all to already look alike normally. In fact, sheep have a very highly developed ability to remember each others facial differences, which indicates that their alikeness is not an ovine perception.

12 The difference between the Dolly technique and the cellular reconstruction that has long gone on in science at an experimental level is its connection to commerce and industry. It could also be called enterprised-up biology, in the same way Marilyn Strathern has written about the enterprising up of nature in the context of new reproductive technologies (1992a, 1992b). Or we could name it de-natured biology, as in a biology that has had its original properties, or nature, modified.

13 For a further discussion of biology as performative, that is, more concerned with what things do than what they are, see Franklin 2005, 2003c, 2001a, and 2000. See also Hayden 2003.

14 PPL Therapeutics, Roslin's commercial partner in the Dolly experiments, was unsuccessful in marketing their first transgenic ruminant-derived product and succumbed to bankruptcy in 2004, as I discuss in chapter 5. The company, whose initials stand for Pharmaceutical Proteins Limited, was initially interested in using sheep to manufacture the proteins missing from sufferers of hemophilia.

15 The kin-sheep system of which Dolly was a part is often invisible in accounts of her creation and her life, although keeping her with other sheep was considered essential to her care. Dolly's paddock mates, including Megan and Morag, are all also part of a network including the sheep who failed to thrive. The presence of death in the sheep family is the subject of my chapter 5.

16 As is often pointed out, Dolly was "icing on the cake" for the proof of the technique used to make her (Wilmut, Campbell, and Tudge 2000, 20). The earlier success of the experiments with sheep cloned from fetal and embryonic cells that had differentiated, Megan and Morag, provided the initial breakthrough ewes.

17 *Totipotency and pluripotency* are increasingly used interchangeably and are defined as the primordial germinal condition of being able to become

any type of tissue. The early embryo, for example, is pluripotent. Other cells, most notably some blood cells and bone marrow cells, are known as progenitor cells because they not only produce other cells like themselves but also cells that are more differentiated.

18 August Weissman's theory of the continuity of the germplasm (Weissman 1892) posited a distinction between so-called germline cells—such as egg and sperm, containing DNA that replicates in perpetuity—and somatic cells, which were, in effect, the mortal containers for this process over time. The idea that body cells can, like eggs and sperm, provide the complete basis for creating new life undermines this distinction and raises a question about the meaning of specialization if, in fact, this is not what occurs.

19 The increasing overlap between binary and genetic codes is evident in the rapidly expanding field of computational biology.

20 The *New York Times* science journalist Natalie Angier describes Wilmut's discovery at the end of the first chapter of her recent Pulitzer Prize-winning book, *Woman: An Intimate Geography* (1999). Giving a new meaning to the term *ovation*, she writes that "the egg proper is the true sun, the light of life, and I say this without exaggeration. The egg is rare in the body and rare in its power. No other cell has the capacity to create the new, to begin with a complement of genes and build an entire being from it. . . . What the sweet ovine face of Dolly demonstrates without equivocation is the wonder of the egg. The egg made the clone. . . . The egg body resurrected the entire adult genome. . . . The egg alone of the body's cells can effect the whole. . . . Eggs must plan the party" (16–17). Angier is possibly unaware that sheep are a strongly matriarchal species, and that the Scottish Blackface ewes who contributed both the egg that made Dolly and her gestational environment are considered to be among the hardiest of sheep. If she did, Angier might be tempted to draw yet another "eggs-aggerated" analogy extolling extreme feminine biological fortitude by claiming that sheep are themselves the ova of the British landscape, their renowned generativity and hardiness providing the equivalent of the egg cell's potency to a nation's economy for centuries.

21 Gene activation is also described as the onset of transcription, the process by which DNA makes RNA (which then makes protein, according to the central dogma of molecular biology coined by Francis Crick in 1963). Transcription, then, is the switching on of the embryonic genome, and the beginning of its action, which changes from being replication (of DNA) to production (of RNA) and thus the messaging or coding function by which

genetic instructions from DNA inside the nucleus begin to be carried out by the actions of other molecules (RNA) and organelles (ribosomes).

2. Capital

1 Thompson's model of the biotech mode of reproduction is discussed further in relation to Marxist models of capital, and the neglected importance of reproduction to capital formation, in Franklin and Lock 2003b, 6–14.

2 The term *breed wealth* is an umbrella category to describe the ways in which reproductive power is reworked to make animals and animals' reproductive power, or value, and is harnessed to accumulate wealth (Franklin 1997a). The classic example would be the work of the master breeder Robert Bakewell in Britain in the mid-eighteenth century. See Ritvo 1995 for an account of Bakewell's New Leicester sheep (also discussed by Marx in *Capital*). For a further discussion of Bakewell's revolutionary breeding techniques in relation to sheep breeding more generally, see Wood and Orel 2001, esp. chapters 4 and 5. For a comparison to Dolly, who constitutes a more recent form of breed wealth, see Franklin 1997a, 2001b, 2001c, 2003b, 2003c, and Franklin, Lury and Stacey 2000, chapter 3.

3 The intended commercial functions of using somatic cell nuclear transfer to create flocks of genetically modified sheep (transgenesis), or to improve the quality of livestock herds by cloning elite animals, have not yet proven to be economically viable.

4 UK patents covering the use of quiescent cells for nuclear transfer were granted in January 2000.

5 Several definitions of *livestock* suggest ownership of reproductive power, but none state this explicitly. The ownership of slaves as livestock has numerous implications for the association of cloning with slavery, most obviously the alienation of reproductive power, and what Hortense Spillers (1987) refers to as the effect of "distributed maternity," a genealogical effect she names as "grammatical" (see also Alys Weinbaum's important 2004 discussion of slavery and alienated reproductive power).

6 See Janet Carsten's (2001) insightful discussion of the etymology of *substance*, its links to ideas of kinship and genealogy, and the suggestive way in which its materialism overlaps with ideas of stock in her chapter entitled "Substantivism, Antisubstantivism, and Anti-antisubstantivism."

7 A particularly important example of the emergence of life stock is human cord blood banking as a form of insurance against future disease.

8 These world-shaping fictions would include the view of human progress as a triumph of scientific rationality, as opposed to the cyborg view

that we embody progress as *irrationality*, for example, in the ways the food chain has toxic effects, or the ways in which health care systems foster inequality.

9 Published in 1983 and by far the most comprehensive ovine-opedia, Ryder's 846-page book contains over 250 sheep-related illustrations drawn from "all possible sources on the history of the association of sheep with man" and compiled in the tradition of the eminent Victorian sheep scholar and veterinarian William Youatt, who, according to Michael Ryder, "exhibited an amazing knowledge of world sheep and their biology" (1983, vii). An expert on the biology of wool in Edinburgh's Animal Breeding Research Organization, a precursor of Roslin, Ryder's scholarship forms an essential background to the making of Dolly the sheep.

10 The opposite of successful domestication is the rogue, stray, or "wild" animal that cannot be tamed. However, such animals are as much products of domestication as their successfully tamed counterparts because it is domestication that makes of their wildness both failure and otherness. The romance attaching to the wild or untamable animal is an example of the anthropocentrism inevitably linked to the very idea of a domestic animal. That the borders of successful domestication are defined by its failure, and indeed require failure, in order for breeding to be selective, stands as an example of the important relationship between biological control and its opposite, what could be called "rogue biology," as I discuss in chapter 5 of this book. Cells, too, are "rogue" when they are, for example, cancerous, and much stem cell research was initiated using cancer cells.

11 For a detailed account of the scientific effort to capture and control the powers of cellular vitality, see Michael D. West's 2003 *The Immortal Cell: One Scientist's Quest to Solve the Mystery of Human Aging*.

12 Other than dogs, sheep are the domestic animals most closely interwoven into human sociality to an extent that, save for Ryder's work has gone virtually without historical commentary (see also Carter 1964, 1979, 1988).

13 Tobias Rees's ethnographic work with stem cell scientists in France, for example, suggests that they rely not only on a model of plastic *biology* but *reason* (personal communication and unpublished papers, 2002–3).

14 *Stem cells* is a confusing term because it refers both to embryonic cells (ES cells, which are taken from early embryonic tissue, usually the ICM) and to undifferentiated progenitor cells derived from specific types of tissue (such as blood and bone marrow). ES cells are considered to be pleuripotent or totipotent, that is, capable of producing any tissue in the body, whereas stem cells taken from blood or marrow are multipotent, meaning

that they can form most, but not all, tissue types. ES cells are the ultimate stem cells, but, in a sense, they are not really stem cells at all, since that term more accurately describes the undifferentiated progenenitor cells that produce specific cell types.

15 The way stem cells are being redefined as productive mechanisms brings to mind comparisons to early industrialization in the north of England, for example, in the way rivers came to be seen as sources of energy and could be redesigned through sluices and weirs to drive waterwheels. Missing from many accounts of the supposedly limitless possibilities of human embryonic cell lines is any attention to the intense demands of their propagation, which is highly laborious. Careful study of the development of stem cell science will require attention to the divisions of labour that come to define the stem cell industry, including, as the case of the disgraced South Korean scientist Woo Suk Hwang has shown, those of provenance and procurement of embryos as well as the production of new lines.

16 The House of Lords would not have appointed a Select Committee to consider the issue of stem cells were it not a matter of significant national concern, while, in keeping with over fifteen years of debate on related matters in Parliament, the outcome of the committee's deliberations is extremely permissive—indeed almost radically liberal. It is a more comprehensive and substantial endorsement of stem cell research than has been produced in any other country (for further discussion of stem cell developments in Britain, see Franklin 2001b, 2003b, and 2005).

17 In early 2004 the British science journal *Nature* published reports of ethical improprieties at South Korea's leading cloning laboratory. These proved to include unethical procurement of eggs as well as scientific and financial fraud. See further Franklin 2006a, 2006b.

18 The press was responding to an article the King's team had submitted to Bob Edwards's journal, *Reproductive Medicine Online*, the abstract of which became available for preview in late July 2003 on the Internet. It was not yet published in full, but the news was out, confirming Britain's much-heralded scientific potential in the stem cell field (Pickering et al., 2003).

19 These meanings of *colony* and their proximity to gardening indicate the breadth of scope available for considering both colonialism and postcolonialism within a biological frame. Since colonialism, in its settler sense, depended heavily on biologicals—such as plants and animals, and also the spread of disease—it is important to consider its spread as more than horticultural, as I discuss further in chapter 4 (and see Crosby 2004, Drayton 2000, and Parry 2004).

20 The reference also indexes the importance of unknown factors in sci-

ence, such as why some lines develop and thrive while others never get off the starting blocks or suddenly cease to thrive. Whereas some of the leading stem cell laboratories in Britain have used hundreds of embryos without creating a single successful cell line, Sue seems to be able to grow her own with exceptional ease.

3. Nation

1 This list is adapted from a longer one compiled by Elspeth Wills in her account of Scottish firsts (2002). The list also includes television, the telephone, the grand piano, the video recorder, absolute zero, and unshrinkable underwear (Wills 2002, 7; see 23–34 about Dolly's kindred firsts in the history of the life sciences in Scotland).

2 For a further discussion of foot and mouth disease, see chapter 5 of this book and also Franklin 2001c.

3 Joseph Banks was president of the Royal Society for over forty years and played a leading role in the history of sheep by introducing Merino sheep to Britain in 1800 (see further in Carter 1964, 1979), but also by, as Patricia Fara (2003) argues in her recent account of Banks, binding together British science, industry, and empire, as epitomized by his role as "Father of Australia" (see Maiden 1909).

4 Britain's Rare Breed Society is actively involved in sheep preservation and, as a result of the foot and mouth crisis discussed further in chapter 5, now operates a sheep biobank. The society also runs an accreditation scheme for butchers to promote the sale of rare-breed meat as a means of strengthening support for the breeds' preservation.

5 The changing fortunes of the distinctive Herdwick sheep have long been used as a bellwether for the changing conditions of the tourist industry in the famous Lakeland region of northern England where the breed was introduced by the author and biologist Beatrix Potter (Lane 1978). Semiferal, and featured increasingly as nouvelle cuisine in trendy London eateries and Lakeland country pubs, the Herdwick epitomizes the ongoing centrality of sheep to evolving definitions of Britishness.

6 The way in which the bloodlines of hardy female mountain sheep are put to economic use to "carry" or "pass" the value of remote, non-arable, and virtually uninhabitable (except by sheep) land through successive crosses offers an example of a traditional system of integrating ecological diversity, seasonal variation, a genealogical calculus of sheep types, and the reproductive cycle into a production line, or a food chain, in ways that reveal the literal convergence of national economies and animal breeding and are deserving of more careful reappraisal in the post-Dolly period.

7 For an excellent account of Britain's distinctive relationship to the life sciences figured by the image of "babies in bottle," see the book of the same name by the American feminist literary critic Susan Squier (1994).
8 The British artist Andy Goldsworthy's "Arch" project retraced these "sheepways," building on his earlier study of sheepfolds, as I discuss further in chapter 5.
9 There are approximately 60 million sheep in Britain, representing more than eighty breeds, more than 80 percent of which are traditional, or ancient, occupants of the British Isles, while approximately 20 percent constitute newcomers.
10 The world's largest sheep producer is in fact China, although Chinese sheep farming is significantly less well developed than that of New Zealand or Australia. See Longworth and Williamson 1993.
11 It is now thought that most of the dominant features of sheep domestication were well established in Asia as many as ten thousand years ago. By 3000 BC, written records and pictorial representations of sheep in Mesopotamian and Babylonian civilizations depicted small, multicolored animals, some of which were polled.
12 Sheep also (1) have no sharp teeth; (2) are usually smaller than humans; (3) have more uses than any other animal; (4) can be herded by dogs; (5) graze more thoroughly than cows, pigs, horses, or deer; and (6) can survive on minimal supplies of food and water.
13 As has frequently been noted throughout this book, the meanings of both domestication and pastoralism are both confirmed and put into question by Dolly's creation and the form of biological control it involved.
14 A sport is a genetic mutation. Sports are central to breeding programs designed to promote specific traits, which can arise through spontaneous mutation and can be transferred and fixed into a breed. Some species are highly mutagenic and consist of hundreds of closely related sports, such as the familiar English Hawthorne, whose varying flower color is but one of its many sportive traits. The highly mutagenic nature of maize was essential to its cultivation into a plant with large ears on its main stem. This is also the reason maize has proven central to the history of both agriculture and genetics, much as have sheep.
15 Fulling was a process of cleaning and thickening cloth by pounding it with mallets—a process semiautomated by the fulling mill in the eleventh century.
16 The story of Joseph Banks's Merino flock has been extensively documented by the sheep and wool historian Harold B. Carter (1964, 1979, 1988), whose argument, like that of Patricia Fara more recently, is that "there can

be little doubt now that it was Sir Joseph Banks who placed the Spanish Merino and the essential knowledge of its breeding, management and productive attributes at the industrial disposal of Great Britain at one of the decisive periods in her industrial evolution" (1979, xvii).

17 Superfine Merino wool could be as thin as 90S. (or 13 to 15 microns).

18 Small numbers of Merino sheep were exported to the New World with Hernando Cortes, who brought them to Mexico in 1530, and with Pedro Menéndez de Avilés to Florida in 1565, accounting for their predominance among the sheep of Native American peoples of the Southwest, such as the Navajo Churro sheep. Small exports of Merino sheep as gifts of the Spanish crown were made to Sweden (1723), Germany (Hanover, 1765) and France (1786) and led to the diversification of Merino types, a process that reached its zenith in Australia in the nineteenth and twentieth centuries. The Napoleonic Wars led to the decline of Merino sheep in Spain, although worldwide they remain the most numerous sheep comprising approximately half of the global sheep population.

19 Bakewell's sheep experiments are legendary, although his breeding methods remain debated. For a stimulating and erudite account of Bakewell's influence on modern genetics, and the relationship of sheep breeding to the work of Gregor Mendel, see Wood and Orel 2001.

20 The Teeswater, for example, was claimed by Trow-Smith to have "become so filled with Dishley blood that it was not easy to find a pure specimen of the breed" (1959, 66).

21 These workers, often displaced subsistence farmers from remote rural areas, were themselves the products of a cycle of land enclosure for which sheep were both cause and cure: the emergence of the proletariat made sheep more profitable, while the very profitability of sheep increased, or fed, the proletariat by displacing more potential recruits from their land.

22 Many Highlanders fled to Britain's fledgling colonies including Australia, where, as the following chapter illustrates, they played a crucial, if ironic, role in displacing the Aboriginal population with sheep.

23 It is often said that the history of Britain is the history of world trade, but in fact it could even more precisely be said that it is the history of wool trade.

24 The battle of Culloden was fought near Inverness in 1746. It marked the Highlanders' last stand to protect their way of life, but it also constituted the last stage in the attempt to restore the Stuart dynasty to the British throne. The Jacobite army, led by Bonnie Prince Charlie, was defeated in the last battle ever fought on mainland British soil.

4. Colony

1 In 1787 Great Britain was still in economic free fall following its army's expensive defeat at the hands of the defiant, and now constitutionally independent, former subjects in the failed colonial project of New England.

2 The sheep forming Australia's first flocks came from many countries including Ireland, India, South Africa, and California. Later they were imported from Spain, Germany, Vermont, and France. The wool industry, not fully established until the 1860s, came eventually to rely on many different types of Merino sheep, among them the famous Peppin Merino, described by some sheep historians as "a triumph of cross-breeding with genes from many types of sheep, including the most primitive" (Garran, xiv).

3 We need not assume the suitability of sheep to Australia, then, to have been in need of particularly onerous fine-tuning or careful adjustment. It may well have been all but self-evident that sheep, like rabbits, would do well there.

4 Significantly, Ker Conway argues, Australia's pastoral glories only began to be achieved after several decades of uneven development and against significant obstacles, including active suppression by the Home and local governments. Contrary to the initial blueprint for the colony's settlement (according to which sheep would be bred for food, not fine wool), and beset by numerous practical obstacles, Australia became a land of sheep not so much deliberately as fortuitously and haphazardly.

5 In 1815 the Australian colonies provided 73,179 pounds of wool to Britain, in comparison to the 6,927,934 pounds Britain imported from Spain. By 1849, this situation was more than reversed. While Spain's wool export to Britain had dropped to 127,559 pounds, Australia's had risen to 35,879,171. By 1888, Australia was home to 97,983,960 sheep.

6 Roberts held the Challis professorship at the University of Sydney, and the first edition of *The Squatting Age in Australia*—long considered a classic contribution to Australian historiography, and part of a tradition established by Collier in the early twentieth century of romanticizing the early squatters and their outback adventures—was published on the centennial of this pivotal date, in 1935. I have used the updated and corrected edition published in 1964.

7 Mrs. Partington is an anecdotal English figure who tries to mop up a tidal wave and thus represents a person waging a futile, hopeless, or hugely disproportionate contest.

8 As Collier later describes the legalization of squatting, it ended the "abortive experiment" of its suppression by ensuring "the ground was left clear for a right design that was of Nature's devising" (1911, 5).

9 This phrase is used in Ralph Trouillot's account of the history of anthropology and critique of the "global geography of imagination" linked to "the West" in which "a space for the inherently Other" is a requisite component (2003, 1–2).

10 It is the separation between black and white Australians that Patrick Wolfe argues is rooted in the frontier effect of "a binary opposition which counterpoised two pure types (civilization vs. savagery, etc.)" (1994a, 95).

11 This use of the genealogical as a reckoning device could be called genealogical orienteering. For a comprehensive discussion of the work of orientations as reckoning devices and modes of creating alignment, see Ahmed 2006.

12 Wolfe's account of the frontier is central to the discussion of colonial frontiers in Lynette Russell's anthology (2001) concerning the legacies of colonial frontiers. See also Bird and Davis 2005 and Slotkin 1973 for a discussion of the "mythology of the American frontier" in the Australian context, and Wolski 2001 for an overview of Australian "contact historiography."

13 According to Watson (1984), the Gippsland Highland Brigade massacred 150 Kurnai people at Warrigal Creek in 1843.

14 Angus McMillan, a leader of his Scottish countrymen and "discoverer of Gippsland," was hailed both in Britain and Australia as "an intrepid explorer, a successful squatter, and a citizen of whom the colony may well feel proud" (Watson 1984, 209). Despite his participation in the Warrigal Creek massacre, he was appointed protector of the Aborigines in the early colonial era, in the pursuit of which role he was praised by the colonial government for his kindliness.

15 Although historians such as Geoffrey Blainey (2001) refer to the Aboriginal people as the masters of the Australian continent, their mode of inhabiting land was neither masterful nor possessive in a Western idiom, and the term *domestication* would be problematically applied to their relationships to animals, soil, plants, or mode of habitation.

16 The frontier is defined, like a horizon, as both an outer limit and as a limitlessness. A conventional distinction between the European idea of the frontier as a fixed, stable, and breachable boundary, and the American, colonial, or New World idea of the frontier as mobile, fluid, and multiple is made by historians such as Fred Alexander (1969).

17 In an important commentary on the frontier hypothesis of Frederick Jackson Turner, the American historian John T. Juricek provides an elaborate dissection of the word "frontier' and its uses, including many which reverse the "toward" of the frontier orientation, and challenge its concep-

tion as a line (1966, and my thanks to Bruce Greenfield for sending me a copy of this extremely useful article). As Juricek notes, the frontier only began to become a "line" in the context of seventeenth century statecraft: "The new national bureaucracies demanded administrative tidiness, including sharply defined frontiers" (1966:13). This shift was also linked to the emergence of the frontier as a singularity, or dividing line, of particular importance to the business of "trad[ing] pieces of territory back and forth at conference tables," he claims (1966, 13).

18 Turner also wrote his essay in direct opposition to the claim made by the superintendent of the census in 1890 that because there no longer existed any frontier line the frontier itself had ceased to exist.

19 According to Ray Allen Billington's introduction to the 1961 edition of Turner's essay, the "negative influence that helped mould the young historian was provided by his Principal Instructor at Johns Hopkins, Professor Herbert Baxter Adams, [whose] sole concern was the evolution of democratic institutions from their 'germs' in Medieval Germany" (1961, 3).

20 Significantly, although somewhat confusingly, the intimacy of contact with native peoples provides one form of union (between settlers and natives), while the "common measures of defence" (against "Indians") establishes a different union (among whites) (Turner 1961, 46).

21 I am indebted to Sara Ahmed for several helpful discussions about this chapter in general and in particular for her insights into the importance of the contact zone as a place of close encounter. For further discussion of close encounters, see Ahmed 2000, especially the introduction and chapter 8.

22 The frontier, a ubiquitous figure in representations of medical, technological, and scientific advancement, is, in this ability to be both someplace and no place, very similar to the idiom of the horizon—also a commonly used idiom for the beyond of imagined future progress, such as the acquisition of new knowledge.

23 McClintock's brilliant and vivid 1995 analysis of the gendered, racial, class, and familial dimensions of colonialism begins with an extensive discussion of the sexualization of the "virgin territory" conquered and charted by white European male explorers through a process she compares to "male birthing." The emphasis on paternity in genealogy owes much to, and tries to combine, what McClintock describes as "sexual" about origins (29), "panoptical" about history (37), and "familial" about time (38). See also Annette Kolodny's analysis of Turner's language as gendered and sexualized (1975, 136–37).

24 Juricek somewhat unfairly criticizes Turner's essay for its author's ahistorical over-reliance on a "unique species of 'frontier'" that was "of little influence" and "embodied a conception that was practically inapplicable for any purposes other than mapmaking and the tabulation of census data" (1966, 29). However, as these passages show, it was precisely Turner's, in his own terms "elastic," use of the frontier concept that so brilliantly captures its ideological potency and enduring centrality to American national culture.

25 Detailed accounts of the application of the American frontier hypothesis to the Australian frontier can be found in Abbott 1971, Alexander 1969, and Perry 1963.

26 According to the lengthy entry on sheep in the *Australian Encyclopedia*, "the crossing of the Blue Mountains in 1813 and the unlocking of the slopes and the plains to the west opened up a new chapter in the sheep industry" (1958, 90), although one could just as easily argue the reverse to hold true.

27 The most recent biography of Macarthur is entitled *Man of Honour* (Duffy 2004).

28 There is less effort to recover lost mothers, or nonwhite fathers, although it should be noted that the Aboriginal scientist David Unaipon, whose numerous inventions include the handheld mechanical sheep-shearing device, appears on the Australian fifty-dollar note, along with Edith Cowan, a campaigner for women's rights.

29 The conspicuously male Merino sheep that stands in the center of the green two-dollar note, facing in the same direction as a twenty-something John Macarthur, features an impressive array of woolly curlicues so thickly coiled that they seem almost deliberately to caricature the luxuriant locks of their "founder" Macarthur. Macarthur is noticeably pushed aside by this magnificent stud, perhaps suggesting the sheep, not the man, represent Australian economic potency.

30 The reference in Graham's account to a belief in "these sheep" as part of a "creed" also evokes Stephen Roberts's (1935) references to the primordial protoplasm of pastoralism, with which Australian settlement is often seen as continuous. Sheep, oddly, in both of these senses become part of a pastoral fundamentalism that provides a quasi-religious legitimation of the "rightness" of occupying indigenous land.

31 Macarthur's statement was supported by evidence from wool manufacturers and was presented to the Committee of the Privy Council on Trade and Foreign Plantations in Lord Hobart's office in London on 26 July.

32 The questions offered by the museum's brochure invite exploration:

"Who were the Macarthurs? How did they manage their estate? What was it like to live here? Where did they fit into colonial society and why is this place important today?"

33 As Jill Ker Conway notes in her account of Elizabeth Macarthur, she "had to supervise a bevy of convict shepherds and their overseer, keep records of sheep numbers, assess the effectiveness of culling methods, decide the number and types of sheep to be sold, direct and evolve suitable methods for sorting, packing and shipping the wool, and prepare returns of all of these activities for her impatient husband in England. Added to this were the full scale agricultural activities of Elizabeth Farm at Parramatta, and the care of a trio of small daughters, one of them in delicate health" (1961, 39).

34 Ker Conway's descriptions of genealogical distance as a source of entrepreneurial advantage deserve further exploration beyond the scope of this chapter, as they introduce a notion of family bond that is explicitly financial, challenging the traditional anthropological assumption of a symbolic distance between "love and money" (Schneider 1968). Moreover, the way this international kinship distance promotes family stock trading is through the assumption that family members share common interests. The forms of symmetry and asymmetry that go into these kinship choreographies would bear significant further scrutiny.

35 John Macarthur's physical and psychological health had been damaged by illness on the onerous original voyage to New South Wales, and his notorious emotional instability was exacerbated by manic depression. Frequently bankrupt, deported, and threatened with court-martial for seriously wounding a superior officer in a duel, Macarthur was confined to his rooms in a state of insanity by the time of his death. He has many detractors who view his esteemed reputation as a farsighted patriot and founder of the sheep-and-wool industry as classic whitewash.

36 This tradition is continued in contemporary popular culture in Australia. For example, the winter 2004 issue of the R. M. Williams–sponsored glossy magazine *Outback* features an article titled "Wool in the Blood" (Austin 2004) tracing the success of the New England, New South Wales Tully family's superfine Merinos through five generations of fathers, and noting in the subhead that "Patriarch Ross Tully can trace the family's sheep connections from its beginnings" (Austin 2004, 46).

37 For insightful discussions of paternity and enterprise, see Coward 1983, McKinnon 2002, and Yanagisako 2002.

38 In an insightful account of family genealogy and money in the context of animal breeding in contemporary British society, the anthropologist

Rebecca Cassidy argues that genealogical connection in the thoroughbred horse racing community of Newmarket, England, not only interdigitates between animal and human bloodlines, but that biological connections are established through an ideology of pedigree linking specific families and individuals to successful animals (Cassidy 2002, 42).

39 Australia celebrated its bicentennial in 1988 (somewhat controversially commemorating the landing of the First Fleet in New South Wales in 1788). The first test-tube baby was born in the United Kingdom in 1978. Australia's first success in this high profile field of medical achievement came two years later in 1980, in Melbourne, making it the third country, behind Britain and the United States, to succeed in this highly competitive field.

40 Central to Trounson's master's thesis, according to Kannegiesser, was the question of "whether multiple births were caused by the ewe producing more eggs than normal or by her uterus accepting additional eggs for implantation" (1988, 342)—a question that was in time to become key to Trounson's work in human IVF.

41 A distinguishing feature of British imperial expansion throughout the eighteenth and nineteenth centuries was its relationship to scientific rationality, which provided both its moral justification in an ethos of "improvement," and its modus operandi drawn from industrial engineering, agriculture, and other branches of applied science (Drayton 2000).

42 As T. J. Robinson, the editor of a volume of papers derived from a six-year study of sheep ovulation induction at the University of Sydney published in 1967, notes, the basic elements of control of sheep ovulation were already well established by 1960, and it was the solution to quite precise problems, such as injection procedures and a "sharp end point" to progesterone treatment, that remained to be discovered in sheep (1967, xiii).

43 Or "scholarsheep," one might say.

44 The patrilineal nature of these legacies deserves further comment than I have provided here, reflecting not simply the gender of science (Jordanova 1989; Keller 1985; Schiebinger 1989) or the sexual politics of biology (Birke 1986; Haraway 1990; Hubbard 1997) but also the distinctive conflation of paternity with scientific genius (Battersby 1989)—a semiotic collapse with very overdetermined implications in the arena of animal breeding (Cassidy 2002; Haraway 2003; Ritvo 1987, 1997).

45 Fittingly, Banks exchanged an Australian kangaroo, among other imperial trophies, for his first Merino sheep, which were illegally imported from France and selectively crossbred with Robert Bakewell's sheep.

5. Death

1 According to the Australian Meat and Livestock Agency, "Australia is the world's largest and most successful exporter of commercial livestock." With 774,358 cattle, 70,913 goats, and 4.7 million sheep exported in 2003, it is clear that this successful export trade is overwhelmingly ovine. See the agency's Web site at www.mla.com.au (accessed 17 April 2004).

2 The "push" market in "off-turns" and "off-cuts" is enabled through cheap transportation on cargo ships, making twenty-first-century global trade in so-called "off-alls" an interesting example of the global mobilities through which "useless" agricultural products are recapitalized.

3 Australia also exports prepacked halal meat to Islamic countries, but the demand is higher for live animals due to the fact that many customers cannot afford the packaged product or do not have refrigeration to keep it fresh.

4 In September 1999, the Australian government announced a 6-million-dollar scheme to promote the improvement of lamb quality, increase demand for lamb, improve lamb processing, and encourage farm productivity and innovation. On 21 December 1999, a press release from the Sheepmeat Council of Australia announcing the launch of the Lamb Industry Development Program (LIDP) commended the government's intention to fast-track the implementation of growth paths for the Australian lamb industry. See www.sheepmeatcouncil.com.au (accessed 11 September 2003).

5 This "westwardization" of sheep shipments out of Australia in fact comprises an "easternization" of markets, signaling some of the changes in global sheep trade that would profit from further analysis in terms of both postcolonial and globalization theory. Australia also supplies live exports to Israel, Singapore, Switzerland, and Malaysia.

6 According to live-export figures from the Australian Meat and Livestock Association, 3,336,846 total head of livestock were exported in 1990, crossing the 4 million mark in the following year, and rising above 5 million in 1993. See www.mla.com.au (accessed 11 September 2003).

7 The National Farmers Association of Australia also paid 10 million dollars in expenses to cover the cost of the animals' care on board the *Cormo* for the additional two months of the voyage.

8 The rising Australian dollar was blamed by Meat and Livestock Australia for the "disastrous fall in sheep exports" in 2003–4, which had dropped by 62 percent in January 2004 (MLA 2004). Since the live-sheep export business deals in otherwise "worthless" sheep, it has a strong supply

drive making it vulnerable in precisely the way the *Cormo* episode demonstrated—that "you can't even give them away." The infection issue could, then, be more of an inflation problem: the epidemic of scabby mouth disguising what was instead an economic casualty. In the sheep trade that provides a vital contribution to the livelihoods of rural Australian farmers, then, the sheep body in transit once again constitutes a complicated vector of market value in livestock, whose survival is purely economic. Thanks very much to Sarah Bell of the faculty of Rural Management at the University of Sydney and University College London for valuable clarification of these and other aspects of the *Cormo* affair.

9 The *Tampa* encountered the stranded twenty-meter fishing boat, the *Palapa*, with 460 mainly Afghan asylum seekers onboard, on 26 August. Three days later, frustrated by official refusals to allow the ship to dock at the nearest port of the Christmas Islands, Captain Arne Rinnan, who was later decorated for his actions by the Norwegian government, entered Australian territorial waters. His ship was forcibly boarded by the Australian military on 29 August, the same day Howard introduced emergency retroactive legislation, the Border Protection Bill 2001. The bill failed to win support from any other party. The refugees were transported aboard an Australian navy vessel to the remote South Pacific island of Nauru, and to New Zealand.

10 In a pro-Howard *Herald Sun* editorial of 8 October 2001, entitled "People We Don't Want," the "children of would-be asylum seekers" were incorrectly claimed to have been "thrown into the sea and used as emotional weapons."

11 Iain Lygo, whose book *News Overboard* offers a detailed analysis of press coverage of the *Tampa* case, describes the anti-Muslim tenor of much Australian media as "Islamaphobic" (2004, 6). See also Sara Ahmed's 2004 discussion of the British "soft touch" discourse concerning refugees and asylum seekers.

12 The power of contagious analogies to establish what Sara Ahmed calls "sticky" connections, for example, in how the "could-be" or "looks-like" terrorist becomes the "might-be" and thus "must-be-treated-as" terrorist, is one of many ways in which contagion can be used as a form of control (Ahmed 2004, 76).

13 A very different view of contagion, namely as a continuum, is central to many non-Western models of disease, as reflected, for example, in the Hindu practice of variolation, a precursor of vaccination based on using small inoculations of cow pox virus to prevent smallpox in humans. For

a brilliant and fascinating comparison of vaccination and variolation, and a critique of the one-way model of infection central to Western medicine and public health, see Apffel-Marglin 1990.

14 Leading authorities on foot and mouth warned in September 2001 of a worldwide epidemic and predicted the epidemic would become global proposing that there should be an international debate on the use of vaccination, instead of culling, to control the disease.

15 Robin McKie, the science editor of the *Sunday Observer*, who originally broke the Dolly story, reported on 9 September 2001 that "Britain's foot and mouth epidemic may have been caused by a cloud of infected dust blown from the Sahara." Dale Griffin of the US Geological Survey and his team based their speculation on live cultures grown from infectious microorganisms that had crossed the Atlantic by clinging to particles of dust (McKie 2001).

16 Argentina, where a foot and mouth epidemic also began in February 2001, had resorted to vaccination by April, whereas in the Netherlands, where the disease also took hold, a belt-and-braces approach was adopted, whereby animals were all vaccinated, but affected animals were also culled. According to the Dutch agricultural minister, Marcello Regunaga, the British approach, of relying entirely on the slaughter policy, was "questionable" from both an economic and an animal welfare standpoint. "I would say that only rich countries could spend the money you need to follow this strategy," Regunaga told the BBC1's *Countryside* program in September 2001 (quoted in Hetherington 2001).

17 In one of the three reports commissioned in the wake of the FMD epidemic entitled the future of farming and food in England, the main theme is "reconnection." As the authors state, "We believe the real reason why the present situation is so dysfunctional is that farming has become detached from the rest of the economy and the environment. . . . The key objective of policy should be to reconnect our food and farming industry: to reconnect farming with its market and the rest of the food chain; to reconnect the food chain and the countryside; and to reconnect customers with what they eat and how it is produced" (Campaign to Protect Rural England 2002, 36).

18 I am grateful to Sara Ahmed for introducing me to the term *affective economies* in her 2004 book *The Cultural Politics of Emotion*, which analyses emotions as effects of circulation.

19 One of the sheep, on the right, is a Scottish Blackface. Like the Cheviots, named for a set of mountains between Scotland and England, the Blackface, or Scottish Mountain sheep, is a hardy northern breed, used for

the "rough grazings" that are at the summit of Britain's strategically tiered sheep-breeding system.

20 In contrast, the *Daily Telegraph* caption began with a single word: "Doomed."

21 The use of the term *ground zero* to describe the epicenter of the epidemic, in conjunction with the repeated imagery of burning pyres of animal bodies and palls of smoke, took on an added set of dimensions in the wake of 9/11. These were not only in the form of catastrophic imagery but also in the way the contagion of FMD became the grounds for increased suspicion and surveillance, and the inevitable scapegoating (or "scapesheeping").

22 The reference to Britain itself as diseased and highly contagious in this editorial is significant and could certainly be read as very pro-European, as well as anti-big agribusiness—a trend significantly exacerbated by Monsanto's GM saga.

23 The Prince of Wales's lucrative organic food company caters to the high end of the market with expensively packaged tea biscuits, oat cakes, and other sundries.

24 MAFF's inefficiency proved terminal, and it was swiftly replaced after the FMD epidemic by a new agency, the Department of Environment, Farming, and Rural Affairs—somewhat inauspiciously known by its acronym DEFRA.

25 So-called fast and processed foods also have this connotation of excessive industrialization or refinement engendering toxicity as well as the taint of over-corporatization.

26 These consequences are now better understood and they are of sufficient concern to scientists such as Wilmut to warrant an international audit of cloned animal health. Studies of the effects of the Dolly technique suggest it may alter delicate mechanisms of genomic imprinting and cellular metabolism, resulting in inborn errors that range from being almost undetectable to being incompatible with life, or so disabling that the animal must be put down.

Breeds

1 The larger project to which Rowell's work belongs is the elucidation of animal sociality, an effort she argues has been impeded by an overemphasis on unifying social systems defined by overvalued structural principles such as hierarchy, or by specific functional requirements such as mating. On the basis of a range of animal studies, Rowell advocates a model of animal sociality in which both individual and group behavioral differences

are assumed to exceed minimum requirements for survival, and to include a range of innovative behaviors neither principally functional nor structural, but expressive of complex animal relationalities. See Rowell 1991b, 1993. My thanks to Thelma for her helpful personal communication and the informative visits to her charming Soay sheep (and goats, turkeys, dogs, and chickens) in the Yorkshire dales.

2 Rowell and her daughter observed a flock of approximately fifty feral Texan Barbados sheep in the Sierra foothills of California. Having been assigned individual numbers, the sheep were observed for an average of five hours per day for five months to generate statistical frequencies of behavior patterns, which were then compared to determine significant correlations.

3 It is beyond the scope of this chapter to address more than superficially the question of whether humor about sheep has changed in the wake of Dolly having been cloned. However it is arguable on at least preliminary anecdotal grounds that they have become more prominent in "knowing" advertisements about "smart" products in Britain, for example a 2002 Carlsberg beer television commercial that plays with the Dolly theme. Increasingly sheep are used in advertising in the print media not only for carpets and wool, but for beer (Youngs brewery 2001), cars (Toyota 2001), pharmaceuticals (Bristol-Myer Squibb 1998), and telecommunications and Internet technology (AT&T 1998, 2000).

4 China has more sheep than any other country.

Bibliography

Abbott, G. J. 1971. *The Pastoral Age in Australia: A Re-examination*. Melbourne: Macmillan Australia.

Ahmed, Sara. 2000. *Strange Encounters: Embodied Others in Post-coloniality*. London: Routledge.

———. 2004. *The Cultural Politics of Emotion*. Edinburgh: Edinburgh University Press.

———. 2006. *Queer Phenomenology: Orientations, Objects, Others*. Durham, NC: Duke University Press.

Alexander, Fred. [1947] 1969. *Moving Frontiers: An American Theme and Its Application to Australian History*. Port Washington, NY: Kennikat Press.

Alexander, Stephanie. 1996. *The Cook's Companion*. Ringwood, Australia: Viking Penguin.

Anderson, K. 1997. "A Walk on the Wild Side: A Critical Geography of Domestication." *Progress in Human Geography* 21, 4: 463–85.

Angier, Natalie. 1999. *Woman: An Intimate Geography*. London: Virago.

Anker, Peder. 2001. *Imperial Ecology: Environmental Order in the British Empire, 1895-1945*. Cambridge, MA: Harvard University Press.

Apffel-Marglin, Frédérique. 1990. "Smallpox in Two Systems of Knowledge." In *Dominating Knowledge: Development, Culture, and Resistance*, ed. Apffel-Marglin and Stephen A. Marglin. Oxford: Clarendon. 102–44.

Apffel-Marglin, Frédérique, and Stephen A. Marglin, eds. 1990. *Dominating Knowledge Development, Culture, and Resistance*. Oxford: Clarendon.

———. 1996. *Decolonizing Knowledge: From Development to Dialogue*. Oxford: Clarendon.

Appadurai, Arjun. 1996. *Modernity at Large: Cultural Dimensions of Globalization*. Minneapolis: University of Minnesota Press.

Ashton, John. 1943. "The Golden Fleece of Early Days: Sheep and Goat History from Pre-historic to Modern Times." *Sheep and Goat Raising*, December: 9–15.

Attwood, Bain. 1989. *The Making of the Aborigines*. Sydney: Allen and Unwin.

———. 2003. *Rights for Aborigines*. Sydney: Allen and Unwin.

Austin, H. B. 1950. *The Merino: Past, Present, and Probable*. Sydney: Grahame.

Austin, Peter. 2004. "Wool In the Blood." *Outback* 35: 46–48.

Australian Meat and Livestock Agency. 2004. "Live Export Markets." (www.mla.com.au, accessed 17 April 2004).

Bailey, Cathy., et al. 2003. "Narratives of Trauma and On-Going Recovery: The 2001 Foot and Mouth Disease Epidemic." *Autobiography* 11, 1/2: 37–45.

Barwick, Sandra. 2001. "Cumbria's Future is Being Slowly Buried." *The Daily Telegraph*. 29 March, p. 14.

Battaglia, Deborra. 2001. "Multiplicities: An Anthropologist's Thoughts on Replicants and Clones in Popular Film." *Critical Inquiry* 27, 3:493–514.

Battersby, Christine. 1989. *Gender and Genius: Toward a Feminist Aesthetics*. Bloomington: Indiana University Press.

Belschner, Henry G. 1971. *Sheep Management and Their Diseases*. 9th ed. London: Angus & Robertson, 1971.

Bhabha, Homi. 1994. *The Location of Culture*. London: Routledge.

Billington, Ray Allen. 1961. "Introduction" to *Frontier and Section: Selected Essays of Frederick Jackson Turner*. Englewood Cliffs: Prentice-Hall. 1–14.

Bird, Deborah, and Richard Davis. 2005. *Dislocating the Frontier: Essaying the Mystique of the Outback*. Canberra: Australian National University Press.

Birke, Lynda. 1986. *Women, Feminism and Biology: the Feminist Challenge*. New York: Methuen.

Blainey, Geoffrey. 2001. *This Land Is All Horizons: Australian Fears and Visions*. Canberra: ABC Books.

Bonnicksen, Andrea L. 2002. *Crafting a Cloning Policy: From Dolly to Stem Cells*. Washington, DC: Georgetown University Press.

Braudel, Fernand. 1979. *The Perspective of the World: Civilization and Capitalism 15th to 18th Centuries, Volume 3*. New York: Harper and Row.

Broadbent, James. 1984. *Elizabeth Farm, Parramatta: A History and Guide*. Sydney: Historic Houses Trust of New South Wales.

Brown, David. 2001. "Phoenix: the Calf They Couldn't Cull, Fights On." *The Daily Telegraph*. 25 April, p. 1.

Bunting, Madeleine. 2001. "Who Are the Brutes Now?" *The Guardian*. 31 March, p. 20.

Burfitt, Charles T. 1913. *History of the Founding of the Wool Industry of Australia*. Sydney: Wiliam Applegate Gullick.

Butler, Judith. 2002. "Is Kinship Always Already Heterosexual?" *Differences* 13, 1: 14–44.

———. 2004. *Precarious Life: the Power of Mourning and Violence*. New York and London: Verso.

Capell, Kerry. 2002. "In Stem Cell Research It's Rule Britannia." *Business Week Online* April 4 (http://www.businessweek.com, accessed 27 April 2002).

Carsten, Janet. 2001. "Substantivism, Antisubstantivism, and Anti-antisubstantivism." In *Relative Values: Reconfiguring Kinship Study*, ed. Sarah Franklin and Susan McKinnon. Durham: Duke University Press. 29–53.

Carter, Harold B. 1964. *His Majesty's Spanish Flock: Sir Joseph Banks and the Merinos of George III of England*. Sydney: Angus and Robertson.

———, ed. 1979. *The Sheep and Wool Correspondence of Sir Joseph Banks 1781–1820*. Norwich, UK: Fletcher.

Carter, Harold. 1988. *Sir Joseph Banks, 1743–1820*. London: British Museum.

Cassidy, Rebecca. 2002. *The Sport of Kings: Kinship, Class, and Thoroughbred Breeding in Newmarket*. Cambridge: Cambridge University Press.

Chisholm, Alan H., ed. 1958. *The Australian Encyclopaedia*. Sydney: Angus and Robertson.

Clutton-Brock, Juliet. 1987. *A Natural History of Domesticated Mammals*. Cambridge: Cambridge University Press.

———, ed. 1989. *The Walking Larder: Patterns of Domestication, Pastoralism, and Predation*. London: Unwin Hyman.

———. 1999. *A Natural History of Domesticated Mammals*. Cambridge: Cambridge University Press.

Collier, James. 1911. *The Pastoral Age in Australasia*. London: Whitcombe and Tomb.

Connor, Steve. 2001. "Sheep Cannot be Accused of Woolly Thinking, Say Scientists." *The Independent Online*. (http://www.hedweb.com .animimag/smartsheep.html, accessed 12 July 2004).

Convery, I. Bailey, C. Mort, and M. Baxter. 2005. "Death in the Wrong Place: Emotional Geographies of the UK 2001 Foot and Mouth Disease Epidemic." *Journal of Rural Studies* 21: 99–109.

Coward, Rosalind. 1983. *Patriarchal Precedents: Sexuality and Social Relations*. London: Routledge and Kegan Paul.

Crosby, Alfred W. 2004. *Ecological Imperialism: The Biological Expansion of Europe, 900–1900*. 2d ed. Cambridge: Cambridge University Press.

Deleuze, Gilles, and Félix Guattari. 1987. *A Thousand Plateaus: Capitalism and Schizophrenia*. Trans. Brian Massumi. London: Athlone.

Denyer, Susan. 1993. *Herwick Sheep Farming: An Illustrated Souvenir*. London: National Trust.

Drayton, Richard. 2000. *Nature's Government: Science, Imperial Britain, and the "Improvement of the World."* New Haven, CT: Yale University Press.

Duffy, Michael. 2004. *Man of Honour: John Macarthur: Duellist, Rebel, Founding Father*. Sydney: Pan Macmillan.

East, Edward M., and Donald F. James. 19191. *Inbreeding and Outbreeding: Their Genetic and Sociological Significance*.

Edwards, Jeanette. 2000. *Born and Bred: Idioms of Kinship and New Reproductive Technologies in England*. Oxford: Oxford University Press.
Engel, Matthew. 2001. "The News From Ground Zero: Foot and Mouth is Winning." *The Guardian*. 17 March, p. 1.
European Commission. 2001. "Stem Cells: Promises and Precautions." *RTD Info* 32 (December): 4–8.
Fara, Patricia. 2003. *Sex, Botany, and Empire: The Story of Carl Linnaeus and Joseph Banks*. Cambridge, Icon.
Feeley-Harnik, Gillian. 1999. " 'Communities of Blood': The Natural History of Kinship in Nineteenth-Century America." *Comparative Studies in Society and History*, 41, 2: 215–262.
———. 2004. "The Geography of Descent." *Proceedings of the British Academy* 125: 311–64.
———. Forthcoming. *The Art of Propagating Life: Charles Darwin and the Pigeon-Breeders of London*.
Fisher, Anthony. 1989. *IVF: The Critical Issues*. Melbourne: Collins Dove.
Franklin, A., and R. White. 2001. "Animals and Modernity: Changing Human-Animal Relations, 1949–98." *Journal of Sociology* 37, 3: 219–38.
Franklin, Sarah. 1995a. "Life." In *Encyclopedia of Bioethics*, ed. Warren Reich. New York: Macmillan. 456–62.
———. 1997a. "Dolly: A New Form of Transgenic Breedwealth." *Environmental Values* 6, 4: 427–37.
———. 1997b. *Embodied Progress: A Cultural Account of Assisted Conception*. London: Routledge.
———. 1999a. Afterword to *Technologies of Procreation: Kinship in the Age of Assisted Conception*, by Jeanette Edwards et al. 2d ed. London: Routledge. 166–71.
———. 1999b. "What We Know and What We Don't about Cloning and Society." *New Genetics and Society* 18, 1: 111–20.
———. 2000. "Life Itself: Global Nature and the Genetic Imaginary." In *Global Nature, Global Culture*, by Franklin, Celia Lury, and Jackie Stacey. London: Sage. 188–227.
———. 2001a. "Biologization Revisited: Kinship Theory in the Context of the New Biologies." In *Relative Values: Reconfiguring Kinship Studies*, ed. Franklin and Susan McKinnon. Durham, NC: Duke University Press. 302–28.
———. 2001b. "Culturing Biology: Cell Lines for the Second Millennium." *Health* 5, 3: 335–54.
———. 2001c. "Sheepwatching." *Anthropology Today* 17, 3: 3–9.
———. 2003a. "Clones and Cloning: New Reproductive Futures." In *Key Issues in Bioethics*, ed. Ralph Levinson and Michael J. Reiss. London: Taylor and Francis. 69–78.

———. 2003b. "Ethical Biocapital: New Strategies of Cell Culture?" In *Remaking Life and Death: Toward an Anthropology of the Biosciences*, ed. Franklin and Margaret Lock. Santa Fe, NM: School of American Research Press. 97–128.

———. 2003c. "Kinship, Genes, and Cloning: Life after Dolly." In *Genetic Nature/Culture: Anthropology and Science beyond the Two-Culture Divide*, ed. Alan H. Goodman, Deborah Heath, and M. Susan Lindee. Berkeley: University of California Press. 95–110.

———. 2003d. "Rethinking Nature/Culture: Anthropology and the New Genetics." *Anthropological Theory* 3, 1: 65–85.

———. 2005. "Stem Cells R Us: Emergent Life Forms and the Global Biological." In *Global Assemblages: Technology, Politics, and Ethics and Anthropological Problems*, ed. Aihwa Ong and Stephen J. Collier. Malden, MA: Blackwell. 59–78.

———. 2006a. "Embryonic Economies: The Double Reproductive Value of Stem Cells." *Biosocieties* 1: 71–90.

———. 2006b. "The IVF-Stem Cell Interface." *International Journal of Surgery* 4, 2: 86–90.

Franklin, Sarah, Celia Lury, and Jackie Stacey. 2000. *Global Nature, Global Culture*. London: Sage.

Franklin, Sarah, and Helena Ragone. 1998a. Introduction to *Reproducing Reproduction: Kinship, Power, and Technological Innovation*, ed. Franklin and Ragone. Philadelphia: University of Pennsylvania Press. 1–14.

———, eds. 1998b. *Reproducing Reproduction: Kinship, Power, and Technological Innovation*. Philadelphia: University of Pennsylvania Press.

Franklin, Sarah, and Margaret Lock. 2003a. "Animation and Cessation: The Remaking of Life and Death." In *Remaking Life and Death: Toward an Anthropology of the Biosciences*, ed. Franklin and Lock. Santa Fe, NM: School of American Research Press. 3–22.

———, eds. 2003b. *Remaking Life and Death: Toward an Anthropology of the Biosciences*. Santa Fe, NM: School of American Research Press.

Franklin, Sarah, and Susan McKinnon. 2001a. Introduction to *Relative Values: Reconfiguring Kinship Studies*, ed. Franklin and McKinnon. Durham, NC: Duke University Press. 1–28.

———, eds. 2001b. *Relative Values: Reconfiguring Kinship Studies*. Durham, NC: Duke University Press.

Fukuyama, Francis. 2002. *Our Posthuman Future: Consequences of the Biotechnology Revolution*. New York: Farrar, Straus and Giroux.

Fuss, Diana. 1996. "Introduction: Human, All Too Human." In *Human, All Too Human*, ed. Fuss. New York: Routledge. 1–8.

Garran, John Cheyne, and Leslie White. 1985. *Merinos, Myths, and Mac-*

arthurs: Australian Graziers and Their Sheep, 1788–1900. Canberra: Australian National University Press.
Gascoigne, John. 1994. *Joseph Banks and the English Enlightenment*. Cambridge: Cambridge University Press.
———. 1998. *Science in the Service of Empire*. Cambridge: Cambridge University Press.
Geist, V. 1971. *Mountain Sheep: A Study in Behaviour and Evolution*. Chicago: University of Chicago Press.
Gilman, Sander. 1986. "Black Bodies, White Bodies: Toward an Iconography of Female Sexuality in Late-Nineteenth Century Art, Medicine, and Literature. In *"Race," Writing, and Difference*, ed. Henry Louis Gates Jr. Chicago: Chicago University Press. 223–61.
Ginsburg, Faye. 1989. *Contested Lives: the Abortion Debate in an American Community*. Berkeley: University of California.
Goldsworthy, Andy, and David Craig. 1999. *Arch*. London: Thames and Hudson.
Graham, Chris. 2000. "Mammalian Development in the UK (1950–1995)." *International Journal of Developmental Biology* 44: 51–55.
Habermas, Jürgen. 2003. *The Future of Human Nature*. Malden, MA: Polity.
Hage, Ghassan. 2003. *Against Paranoid Nationalism: Searching for Hope in a Shrinking Society*. Pluto Press: Sydney.
Hale, John, ed. 1950. *Settlers: Being Extracts from the Journals and Letters of Early Colonists in Canada, Australia, South Africa, and New Zealand*. London: Faber and Faber.
Haraway, Donna. 1989. *Primate Visions: Gender, Race, and Nature in the World of Modern Science*. New York: Routledge.
———. 1991. *Simians, Cyborgs, and Women: The Reinvention of Nature*. London: Free Association Books.
———. 1997. *Modest_Witness@Second_Millennium. FemaleMan_Meets_Onco Mouse: Feminism and Technoscience*. New York: Routledge.
———. 2003. *The Companion Species Manifesto*. Chicago: Prickly Paradigm.
Harris, John. 2004. *On Cloning*. London: Routledge.
Hawkesworth, Alfred. 1930. *Australasian Sheep and Wool: A Practical and Theoretical Treatise*. 6th ed. Sydney: William Brook.
Hayden, Corinne P. 1995. "Gender, Genetics and Generation: Reformulating Biology in Lesbian Kinship." *Cultural Anthropology* 10, 1: 41–63.
———. 2003. "Suspended Animation: A Brine Shrimp Essay." In *Remaking Life and Death: Toward an Anthropology of the Biosciences*, ed. Sarah Franklin and Margaret Lock. Santa Fe, NM: School of American Research Press. 193–226.
Hetherington, Peter. 2001. "Argentina and Netherlands Use Jabs to Con-

trol Disease." *Guardian Unlimited.* (http://www.guardian.co.uk/footand mouth/story/0,7369,548659,00.html, accessed 24 September 2001).
Historic Houses Trust of New South Wales. 1995. *Elizabeth Farm Parramatta: A History and a Guide*. Rev., ed.
Hogle, Linda. 2003. "Life/Time Warranty: Rechargeable Cells and Extendable Lives." In *Remaking Life and Death: Toward an Anthropology of the Biosciences*, ed. Sarah Franklin and Margaret Lock. Santa Fe, NM: School of American Research Press. 61–96.
Horlacher, Levi Jackson. 1927. *Sheep Production*. New York: McGraw-Hill.
Ingold, Tim. 1988. ed. *What Is an Animal?* London: Unwin Hyman.
Jardine, Nicholas, James Secord, and Emma Spary, eds. 1996. *Cultures of Natural History*. Cambridge: Cambridge University Press.
Jeffrey, George. 1910. *Sheep and Wool for the Farmer: A Practical Handbook*. Melbourne: E. W. Cole.
Jordanova, Ludmilla. 1989. *Sexual Visions: Images of Gender in Science and Medicine between the Eighteenth and Twentieth Centuries*. New York: Harvester Wheatsheaf.
Kannegiesser, Harry. 1988. *Conception in the Test-Tube: How Australia Leads the World*. Melbourne: Macmillan.
Kass, Leon. 1998. "The Wisdom of Repugnance." In Leon Kass and James Wilson, *The Ethics of Human Cloning*. Washington: American Enterprise Institute Press. 3–60.
Keller, Evelyn Fox. 1985. *Reflections on Gender and Science*. New Haven: Yale University Press.
———. 2000. *The Century of the Gene*. Cambridge, MA: Harvard University Press.
Kendrick, K.M., et al. 2001. "Sheep Don't Forget a Face." *Nature* 414: 165–66.
Ker, Jill. 1960. "Merchants and Merinos." *R.A.H.S. Journal* 46, 4: 206–23.
———. 1961. "The Wool Industry in New South Wales, 1803–1830." *Bulletin* 1, 3: 28–49.
———. 1962. "The Wool Industry in New South Wales, 1803–30, Part II." *Business Archives and History* II, 1: 18–54.
Kinsella, John. 2001. *The Hierarchy of Sheep*. Freemantle: Freemantle Arts Centre Press.
Kinsman, David. 2001. *Black Sheep of Windermere: A History of the St Kilda or Hebridean Sheep*. Cumbria: Windy Hill.
Klotzko, Arlene Judith. 2004. *A Clone of Your Own? The Science and Ethics of Cloning*. Oxford: Oxford University Press.
Kolodny, Annette. 1975. *The Lay of the Land: Metaphor as Experience and History in American Life and Letters*. Chapel Hill: University of North Carolina Press.

Kolata, Gina. 1997. Clone: The Road to Dolly and the Path Ahead. London: Allen Lane.

———, ed. 2001. *The Cloning Sourcebook*. Oxford: Oxford University Press.

Landecker, Hannah. 2000. "Immortality, in Vitro: A History of the HeLa Cell Line." In *Biotechnology and Culture: Bodies, Anxieties, Ethics*, ed. Paul E. Brodwin. Bloomington: Indiana University Press. 53–74.

———. 2003. "On Beginning and Ending with Apoptosis: Cell Death and Biomedicine." In *Remaking Life and Death: Toward an Anthropology of the Biosciences*, ed. Sarah Franklin and Margaret Lock. Santa Fe, NM: School of American Research Press. 23–60.

Latour, Bruno. 1987. *Science in Action: How to Follow Scientists and Engineers through Society*. Cambridge, MA: Harvard University Press.

———. 1993. *We Have Never Been Modern*. Trans. Catherine Porter. Cambridge, MA: Harvard University Press.

Law, John, and Annemarie Mol, eds. 2002. *Complexities: Social Studies of Knowledge Practices*. Durham, NC: Duke University Press.

Layne, Linda, ed. 1999. *Transformative Motherhood: On Giving and Getting in a Consumer Culture*. New York: New York University Press.

Leach, Helen M. 2003. "Human Domestication Reconsidered." *Current Anthropology* 44, 3: 349–68.

Lévi-Strauss, Claude. 1972. *The Savage Mind*. London: Weidenfeld and Nicolson.

Lock, Margaret. 2001. "The Alienation of Body Tissue and the Biopolitics of Immortalized Cell Lines." *Body and Society* 7, 2–3: 63–91.

———. 2002. *Twice Dead: Organ Transplants and the Reinvention of Death*. Berkeley: University of California Press.

Lydekker, R. 1913. *The Sheep and Its Cousins*. New York: Dutton.

Lygo, Iain. *News Overboard: The Tabloid Media, Race Politics, and Islam*. Sydney: Southerly Change Publications.

Lyotard, Jean-François. 1991. *The Inhuman: Reflections on Time*. Trans. Geoffrey Bennington and Rachel Bowlby. Malden, MA: Polity.

MacKenzie, John M. 1988. *The Empire of Nature: Hunting, Conservation, and British Imperialism*. Manchester: Manchester University Press.

Maiden, J. H. 1909. *Sir Joseph Banks: The "Father of Australia."* Sydney: William Applegate Gullick.

Marglin, Frédérique Apffel. 1990. "Smallpox in Two Systems of Knowledge." In *Dominating Knowledge: Development, Culture and Resistance*, ed. F.A. Marglin and S.A. Marglin. Oxford: Clarendon.

Martin, Emily. 1991. "The Egg and the Sperm: How Science Has Constructed a Romance Based on Stereotypical Male-Female Roles." *Signs: Journal of Women in Culture and Society* 16, 3: 485–501.

———. 1995. "Working Cross the Human-Other Divide." In *Reinventing*

Biology: Respect for Life and the Creation of Knowledge, ed. Linda Birke and Ruth Hubbard. Bloomington: Indiana University Press. 261–75.

Marx, Karl. 1853. "The Duches of Sutherland and Slavery." *The People's Paper* 45.

———. 1965a. *The German Ideology*. London: Lawrence and Wishart.

———. 1965b. *Pre-Capitalist Economic Formations*. Trans. Jack Cohen. Ed. Eric Hobsbawn. New York: New International Press.

———. [1894] 1972. *Capital: A Critique of Political Economy*. Vol. 3. Ed. Frederick Engels. London: Lawrence and Wishart.

McClintock, Anne. 1995. *Imperial Leather: Race, Gender, and Sexuality in the Colonial Context*. New York: Routledge.

McIvor, Clarence. 1893. *The History and Development of Sheep Farming from Antiquity to Modern Times*. Sydney: Tilghman and Barnett.

McKibben, Bill. 2003. *Enough: Genetic Engineering and the End of Human Nature*. London: Bloomsbury.

McKinnon, Susan. 2001. "The Economies in Kinship and the Paternity of Culture: Origin Stories in Kinship Theory." In *Relative Values: Reconfiguring Kinship Study*, ed. Sarah Franklin and Susan McKinnon. Durham: Duke University Press, pp. 277–301.

Melville, Elinor G. K. 1994. *A Plague of Sheep: Environmental Consequences of the Conquest of Mexico*. Cambridge: Cambridge University Press.

Metherell, Mark. 2003. "Sheep Ship Unwanted, Even Back in the Bush." *Sydney Morning Herald*, 15 November.

Miller, David Philip and Peter Hanns Reill, eds. 1996. *Visions of Empire: Voyages, Botany, and Representations of Empire*. Cambridge: Cambridge University Press.

Mol, Annemarie. 2002. *The Body Multiple: Ontology in Medical Practice*. Durham, NC: Duke University Press.

Mullin, M. H. 1999. "Mirrors and Windows: Sociocultural Studies of Human-Animal Relationships." *Annual Review of Anthropology* 28: 201–24.

Nash, Catherine. 2002. "Genealogical Identities." *Environment and Planning* 20, 1: 27–52.

———. 2004. "Genetic Kinship." *Cultural Studies* 18, 1: 1–34.

Parry, Bronwyn. 2004. *Trading the Genome: Investigating the Commodification of Bio-information*. New York: Columbia University Press.

People for Ethical Treatment of Animals (PETA). 2001. "Scientists Say That Sheep Have feelings Too." (http://www.woolisbaad.com/f-feelings.asp, accessed 12 July 2004).

Pence, Gregory E., ed. 1998. *Flesh of My Flesh: The Ethics of Cloning Humans; A Reader*. Lanham, MD: Rowman and Littlefield.

Perry, T. M. 1963. *Australia's First Frontier: The Spread of Settlement in New South Wales, 1788–1829*. Melbourne: Melbourne University Press.

Philo, C., and J. Wolche. 1998. "Through the Geographical Looking Glass: Space, Place, and Society-Animal Relations." *Society and Animals* 6, 2: 103–18.

Pickering, Susan, Braude, Peter, Patel, Minal, Burns, Chris J., Trussler, Jane, Bolton, Virginia, Minger, Stephen. 2003. "Preimplantation Genetic Diagnosis as a Novel Source of Embryos for Stem Cell Research," *Reproductive BioMedicine Online* 7, 3: 353–64.

Poulter, Sean. 2001. " 'This Epidemic is Out of Control'." *Daily Mail*. 24 March, pp. 10–11.

Povinelli, Elizabeth. 2002. "Notes on Gridlock: Genealogy, Intimacy, Sexuality." *Public Culture* 14, 1: 215–38.

Pratt, Mary Louise. 1992. *Imperial Eyes: Travel Writing and Transculturation*. London: Routledge.

Prebble, John. 1963. *The Highland Clearances*. Harmondsworth: Penguin.

Probyn, Fiona. 2003. "The White Father: Paternalism, Denial and Community," *Cultural Studies Review* 9: 1.

Rabinow, Paul. 1996a. *Essays on the Anthropology of Reason*. Princeton, NJ: Princeton University Press.

———. 1996b. *Making PCR: A Story of Biotechnology*. Chicago: University of Chicago Press.

———. 1999. *French DNA: Trouble in Purgatory*. Chicago: University of Chicago Press.

———. 2003. *Anthropos Today: Reflections on Modern Equipment*. Princeton, NJ: Princeton University Press.

Rader, Karen A. 2004. *Making Mice: Standardizing Animals for American Biomedical Research, 1900–1955*. Princeton, NJ: Princeton University Press.

Ragone, Helena. 1994. *Conception in the Heart: Surrogate Motherhood in America*. Boulder: Westview.

Randall, Henry Stephens. 1863. *Fine Wool Sheep Husbandry*. New York: O. Judd.

Rapp, Rayna. 1999. *Testing Women, Testing the Fetus: The Social Impact of Amnoiocentesis in America*. New York: Routledge.

———. 2003. "Cell Life and Death, Child Life and Death: Genomic Horizons, Genetic Diseases, Family Stories." In *Remaking Life and Death: Toward an Anthropology of the Biosciences*, ed. Sarah Franklin and Margaret Lock. Santa Fe, NM: School of American Research Press. 129–64.

Reynolds, Henry. 1982. *The Other Side of the Frontier: Aboriginal Resistance to the European Invasion of Australia*. Harmondsworth: Penguin.

Rhind, S., et al. "Dolly: A Final Report." *Reproduction, Fertility and Development* 16, 2: 156.

Ritvo, Harriet. 1987. *The Animal Estate: The English and Other Creatures in the Victorian Age*. Cambridge, MA: Harvard University Press.

———. 1995. "Possessing Mother Nature: Genetic Capital in Eighteenth-Century Britain." In *Early Modern Conceptions of Property*, ed. John Brewer and Susan Staves. London: Routledge. 413–26.

———. 1996. "Barring the Cross: Miscegenation and Purity in Eighteenth and Nineteenth Century Britain." In *Human, All Too Human*, ed. Diana Fuss. New York: Routledge. 37–58.

———. 1997. *The Platypus and the Mermaid, and Other Figments of the Classifying Imagination*. Cambridge, MA: Harvard University Press.

Roberts, Stephen Henry. [1935] 1964. *The Squatting Age in Australia, 1835–1847*. Melbourne: Melbourne University Press.

Robinson, T. J. R., ed. 1967. *The Control of the Ovarian Cycle in the Sheep*. Sydney: Sydney University Press.

Rorvik, David M. 1978. *In His Image: The Cloning of a Man*. Philadelphia: Lippincott.

Rowell, T. E. 1991a. "Till Death Do Us Part: Long-lasting Bonds Between Ewes and their Daughters." *Animal Behaviour* 42: 681–82.

———. 1991b. "What Can We Say About Social Structure?" In *The Development and Integration of Behaviour*, ed. P. G. B. Bateson. Cambridge: Cambridge University Press. 255–270.

———. 1993. "Reification of Social Systems." *Evolutionary Anthropology*. 135–37.

Rowell, T. E. and C. A. Rowell. 1993. "The Social Organization of Feral *Ovis Aries* Ram Groups in the Pre-Rut Period." *Ethology* 95: 213–32.

Russell, Lynette, ed. 2004. *Colonial Frontiers: Indigenous-European Encounters in Settler Societies*. Manchester: Manchester University Press.

Russell, Nicholas. 1986. *Like Engend'ring Like: Heredity and Animal Breeding in Early Modern England*. Cambridge: Cambridge University Press.

Ryan, L. D. 1973. *Sheep in Australia*. Sydney: Angus and Robertson.

Ryder, M. L. 1983. *Sheep and Man*. London: Duckworth.

Ryder, M. L., and S. K. Stephenson. 1968. *Wool Growth*. London: Academic Press.

Said, Edward W. 1978. *Orientalism*. New York: Pantheon.

Schneider, David M. 1968. *American Kinship: A Cultural Account*. Chicago: University of Chicago.

Schwartz, Hillel. 1996. *The Culture of the Copy: Striking Likenesses, Unreasonable Facsimiles*. New York: Zone Books.

Scott, John, and Charles Scott. 1888. *Blackfaced Sheep: Their History Distribution and Improvement*. Edinburgh: Grange Publishing Works.

Shamblott M., et al. 1998. "Derivation of Pluripotent Stem Cells from Cultured Human Primordial Germ Cells." *Proceedings of the National Academy of Sciences* 95: 13726–31.

Shanklin, E. 1985. "Sustenance and Symbol: Anthropological Studies of Domesticated Animals." *Annual Review of Anthropology* 14: 375–403.

Silver, Lee M. 1997. *Remaking Eden: Cloning and Beyond in a Brave New World*. New York: Avon.

Slotkin, R. 1973. *Regeneration through Violence: The Mythology of the American Frontier, 1600–1860*. Middletown, CT: Wesleyan University Press.

Sofia, Zoe. 1984. "Exterminating Fetuses: Abortion, Disarmament, and the Sexo-Semiotics of Extraterrestialism." *Diacritics* 14, 2: 47–59.

Spillers, Hortense. 1987. "Mama's Baby, Papa's Maybe: An American Grammar Book." *Diacritics* 17, 2: 65–81.

Squier, Susan. 2000. "Life and Death at Strangeways: The Tissue Culture Point of View." In *Biotechnology and Culture: Bodies, Anxieties, Ethics*, ed. Paul E. Brodwin. Bloomington: Indiana University Press. 27–52.

Stacey, Jackie. 2005. "Masculinity, Masquerade, and Genetic Impersonation: Gattaca's Queer Visions." *Signs: Journal of Women in Culture and Society* 30, 3: 1851–77.

Stacey, Jackie. Forthcoming. *The Cinematic Life of the Gene*. Durham, NC: Duke University Press.

Stephens, Philip. 2001. "Britain's Epidemic of Hysteria." *Financial Times*, 20 March, p. 21.

Strathern, Marilyn. 1988. *The Gender of the Gift: Problems with Women and Problems with Society in Melanesia*. Berkeley: University of California Press.

———. 1992a. *After Nature: English Kinship in the Late Twentieth Century*. Cambridge: Cambridge University Press.

———. 1992b. *Reproducing the Future: Essays on Anthropology, Kinship and the New Reproductive Technologies*. New York: Routledge.

———. 1995. "Displacing Knowledge: Technology and the Consequences for Kinship." In *Conceiving the New World Order: The Global Politics of Reproduction*, ed. Faye Ginsburg and Rayna Rapp. Berkeley: University of California Press. 346–68.

———. 1999a. *Property, Substance and Effect: Anthropological Essays on Persons and Things*. London: Athlone.

———. 1999b. "Introduction: A Question of Context." In Jeanette Edwards, Sarah Franklin, Eric Hirsch, Frances Price, and Marilyn Strathern, *Technologies of Procreation: Kinship in the Age of Assisted Conception*. London: Routledge. 9–28.

Thomas, J. F. H. 1955. *Sheep*. London: Faber and Faber.

Thomas, Keith. 1983. *Man and the Natural World: Changing Attitudes in England, 1500–1800*. London: Allen Lane.

Thomas, Nicholas. 1991. *Entangled Objects: Exchange, Material Culture, and Colonialism in the Pacific*. Cambridge: Harvard University Press.

Thompson, Charis. 2005. *Making Parents: The Ontological Choreography of Reproductive Technologies*. Cambridge, MA: MIT Press.

Thomson, J., et al. 1998. "Embryonic Stem Cells Derived from Human Blastocysts." *Science* 282: 1145–47.

Trivedi, Bijal P. 2001. "Sheep Are Highly Adept at Recognising Faces, Study Shows." *National Geographic Today Online* (http://news.National geographic.com/news/2001/11/1107_Tvsheep.html, accessed 12 July 2004).

Trouillot, Michel-Rolph. 1991. "Anthropology and the Savage Slot." In *Recapturing Anthropology: Working in the Present*, ed. Richard Fox. Santa Fe: School of American Research Press. 21–36.

———. 2003. *Global Transformations: Anthropology and the Modern World*. New York: Palgrave/St. Martin's/Macmillan.

Trow-Smith, Robert. 1957. *A History of British Livestock to 1700*. London: Routledge and Kegan Paul.

———. 1959. *A History of British Livestock 1700 to the Present*. London: Routledge and Kegan Paul.

Turner, Frederick Jackson. 1947. *The Frontier in American History*. New York: Henry Holt.

———. 1961. *Frontier and Section: Selected Essays of Frederick Jackson Turner*. Englewood Cliffs, NJ: Prentice-Hall.

Turney, Jon. 1998. *Frankenstein's Footsteps: Science, Genetics, and Popular Culture*. New Haven, CT: Yale University Press.

United States Department of Agriculture Foreign Agricultural Service. 2004. "Sheep." FAS Online (http://www.fas.usda.gov/dlp2/circular/1999/99-10LP/sheep.htm, accessed 2 March 2004).

Urry, John. 1999. *Sociology beyond Societies: Mobilities for the Twenty First Century*. London: Routledge.

Vidal, John. 2001. "Cattle Low Mightily, Then the Thuds Begin." The *Guardian*. 31 March, p. 4.

Walker, J. F. 1942. *Breeds of Sheep*. Chicago: Breeder Publications.

Ward, Russel. 1978. *The Australian Legend*. Melbourne: Oxford University Press.

Watson, Don. 1984. *Caledonia Australis: Scottish Highlanders on the Frontier of Australia*. Sydney: Collins.

Weinbaum, Alys Eve. 2004. *Wayward Reproductions: Genealogies of Race and Nation in Transatlantic Modern Thought*. Durham, NC: Duke University Press.

Weismann, A. 1892. *Das Keimplasma: Eine Theorie der Vererburg*. Jena: Gustav Fischer.

West, Michael D. 2003. *The Immortal Cell: One Scientist's Quest to Solve the Mystery of Human Ageing*. New York: Doubleday.
Williams, Raymond. *The Sociology of Culture*. New York: Schocken, 1982.
Willis, Roy, ed. 1990. *Signifying Animals: Human Meaning in the Natural World*. London: Unwin Hyman.
Wills, Elspeth. 2002. *Scottish Firsts: A Celebration of Innovation and Achievement*. Edinburgh: Mainstream.
Wilmut, Ian, et al. 1997. "Viable Offspring Derived from Fetal and Mammalian Cells." *Nature* 385: 811–13.
Wilmut, Ian, Keith Campbell, and Colin Tudge. 2000. *The Second Creation: The Age of the Biological Control by the Scientists Who Cloned Dolly*. London: Headline.
Wolch, J., and J. Emel. 1998. *Animal Geographies: Place, Politics, and Identity in the Nature-Culture Borderlands*. London: Verso.
Wolfe, Patrick. 1994a. " 'White Man's Flour': Doctrines of Virgin Birth in Evolutionist Ethnogenetics and Australian State Formation." *History and Anthropology* 8, 4: 165–205.
Wolfe, Patrick. 1994b. "Nation and Miscegenation: Discursive Continuity in the Post-Mabo Era." *Social Analysis* 36: 93–152.
Wolfe, Patrick. 1999. *Settler Colonialism and the Transformation of Anthropology: the Politics and Poetics of an Ethnographic Event*. London: Cassell.
Wood, Roger J., and Vitezslav Orel. 2001. *Genetic Prehistory in Selective Breeding: A Prelude to Mendel*. Oxford: Oxford University Press.
Wrightson, John. 1895. *Sheep: Breed and Management*. 2d ed. London: Vinton.
Yanagisako, Sylvia. 2002. *Producing Capital: Family Firms in Italy*. Princeton: Princeton University Press.
Yanagisako, Sylvia, and Carol Delaney, eds. 1995. *Naturalizing Power: Essays in FeministCultural Analysis*. New York: Routledge.
Youatt, William. 1894. *Sheep: Their Breeds, Management, and Diseases*. London: Simpkin, Marshall, Hamilton, Kent.

Index

SARAH FRANKLIN is a professor of social studies of biomedicine at the London School of Economics and associate director of the BIOS centre for the study of bioscience, biomedicine, biotechnology and society. She is the author of *Embodied Progress: A Cultural Account of Assisted Conception* and the coauthor of *Global Nature, Global Culture* (with Celia Lury and Jackie Stacey), *Technologies of Procreation: Kinship in the Age of Assisted Conception* (with Jeanette Edwards, Eric Hirsch, Frances Price, and Marilyn Strathern), and *Born and Made: An Ethnography of Preimplantation Genetic Diagnosis* (with Celia Roberts). She is the editor of *The Sociology of Gender* and the coeditor of *Off-Centre: Feminism and Cultural Studies* (with Jackie Stacey and Celia Lury), *Reproducing Reproduction: Kinship, Power, and Technological Innovation* (with Helena Ragone), *Relative Values: Reconfiguring Kinship Study* (with Susan McKinnon), and *Remaking Life and Death: Toward an Anthropology of the Biosciences* (with Margaret Lock).

Library of Congress Cataloging-in-Publication Data
Franklin, Sarah
Dolly mixtures : the remaking of genealogy / Sarah Franklin.
p. ; cm.
A John Hope Franklin Center book.
Includes bibliographical references and index.
ISBN-13: 978-0-8223-3903-8 (cloth : alk. paper)
ISBN-13: 978-0-8223-3920-5 (pbk. : alk. paper)
1. Dolly (Sheep) 2. Transgenic animals—Social aspects. 3. Cloning—Social aspects. 4. Genealogy. I. Title.
[DNLM: 1. Cloning, Organism—history. 2. Cloning, Organism—economics. 3. History, 20th Century. QU 11.1 F834d 2007]
QH442.6.F73 2007
636.3'0821—dc22 2006032814